T0318634

HANDBOOK FOR LABORATORY SAFETY

HANDBOOK FOR LABORATORY SAFETY

BENJAMIN R. SVEINBJORNSSON

Assistant Professor, Faculty of Physical Sciences, University of Iceland, Iceland

SVEINBJORN GIZURARSON

Professor, Faculty of Pharmaceutical Science, University of Iceland, Iceland

ELSEVIER

Elsevier
Radarweg 29, PO Box 211, 1000 AE Amsterdam, Netherlands
The Boulevard, Langford Lane, Kidlington, Oxford OX5 1GB, United Kingdom
50 Hampshire Street, 5th Floor, Cambridge, MA 02139, United States

Notices
Knowledge and best practice in this field are constantly changing. As new research and experience broaden our understanding, changes in research methods, professional practices, or medical treatment may become necessary.

Practitioners and researchers must always rely on their own experience and knowledge in evaluating and using any information, methods, compounds, or experiments described herein. In using such information or methods they should be mindful of their own safety and the safety of others, including parties for whom they have a professional responsibility.

To the fullest extent of the law, neither the Publisher nor the authors, contributors, or editors, assume any liability for any injury and/or damage to persons or property as a matter of products liability, negligence or otherwise, or from any use or operation of any methods, products, instructions, or ideas contained in the material herein.

Library of Congress Cataloging-in-Publication Data
A catalog record for this book is available from the Library of Congress

British Library Cataloguing-in-Publication Data
A catalogue record for this book is available from the British Library

ISBN: 978-0-323-99320-3

For information on all Elsevier publications visit our website at
https://www.elsevier.com/books-and-journals

Publisher: Candice Janco
Acquisitions Editor: Anita Koch
Editorial Project Manager: Bernadine A. Miralles
Production Project Manager: Paul Prasad Chandramohan
Cover Designer: Miles Hitchen

Working together
to grow libraries in
developing countries

www.elsevier.com • www.bookaid.org

Typeset by TNQ Technologies

To
Kristín Linda Ragnarsdóttir
Davíð Örn Sveinbjörnsson
Guðlaug María Sveinbjörnsdóttir
for continuously inspiring us, believing in us
and reminding us how much we value them.

Contents

Foreword

The most important thing when working in a laboratory is safety—both our own and our coworkers. Unfortunately, accidents can and do happen, but by practicing awareness of the potential hazards involved in our experiments and work environment, we are better able to prevent accidents and respond in such a manner that minimizes any damage of accidents when they do happen.

This book is intended to raise awareness of various safety considerations and provide guidelines for work practices that the authors have found to help enhance the safety in the laboratory. It is impossible to provide guidelines for every situation imaginable or provide guaranteed methods to prevent or react to any type of accident, so it is important to be aware of that while reading this book. While this book provides helpful guidelines for various situations, it is ultimately the responsibility of the individual researcher to make the risk assessments needed and take the appropriate precautionary steps to minimize any danger of their experiments. It is likewise the responsibility of the employer to provide proper safety training for their employees.

At the end of the day, we all want to get home safe and sound, and it is therefore important that we all take the steps that are within our reach to enhance the safety of our work environment and prioritize safety first!

March 17th, 2022
Benjamín Ragnar Sveinbjörnsson
Sveinbjörn Gizurarson

Acknowledgments

In 2010, Sveinbjörn Gizurarson wrote a handbook for laboratory safety in Icelandic to meet the need for a practical, easy to access text on the subject for both students and employees at the University of Iceland. Since then, there has been a regular request to provide the material in English. Benjamin Ragnar Sveinbjornsson joined in on that work and together we built on the original Icelandic handbook. All chapters have been revised, improved, and expanded since the Icelandic version was made. Additionally, we have added real life examples and exercises, where appropriate, to emphasize the importance of the subjects and a new chapter on occupational health, to provide a new handbook that can be used to meet the needs of a wider audience.

A lot of the feedback Sveinbjorn had received on the Icelandic handbook remained useful for the writing of this book; although the feedback is from 2010, we would like to thank the following people for their review and comments on various topics (some have left their positions since then): Sigríður Björnsdóttir (former safety committee employee for the University of Iceland), Sigurður Emil Pálsson (Icelandic Radiation Safety Authority), Kristinn Tómasson, Leifur Gústafsson, and Guðmundur I Kjerúlf (Administration of Occupational Safety and Health in Iceland), Elín Soffía Ólafsdóttir (Dean, Faculty of Pharmaceutical Sciences, University of Iceland), Örvar Aðalsteinsson (The Greater Reykjavik Fire and Rescue Service), Ágúst Svansson (Reykjavik Police Department), Jón Baldursson (former head of ER at the University Hospital, Iceland), Guðborg Auður Guðjónsdóttir (former chair of University Hospital Poison Center, Iceland), Haukur R. Magnússon (The Environment Agency of Iceland), and Eggert Eggertsson (at AGA-ISAGA ehf). Ólafur Ingi Jónsson, who was a good friend who passed away in 2018, worked for the Sudurnes Fire and Rescue Service and was the promoter in establishing fire extinguishing training at the University of Iceland for over a decade. Additionally, we would like to thank Gísli Jónsson (Icelandic Radiation Safety Authority) for additional feedback on new examples.

We would like to thank Elsevier and everyone there who has worked on this book with us, especially Paul Prasad Chandramohan, Bernadine A. Miralles, Cza Osuyos, and Anita Koch. Thank you for helping us make this publication into a reality and for all the work that has gone into making it into the book that it is today.

We also want to thank all those that have assisted us with the figures in this book. We are especially thankful to Birkir Eyþór Ásgeirsson for taking many of the pictures used in this book. We are also thankful for all of those that posed for the pictures. Furthermore, we would like to thank the following individuals and organizations for allowing us to use their material in this book: the World Health Organization (WHO), Helga Sigrún Harðardóttir, Managing Director for the Icelandic Standards (International Organization for Standardization (ISO)), United Nations GHS (UN Globally Harmonized System of Classification and Labelling of Chemicals), Chem-Lab NV in Belgium, Birgir Finnsson (from The Greater Reykjavik Fire And Rescue Service), and Dr. Phatsawee Jansook (Chulalongkorn University, Thailand).

Last, but not least, we want to thank our family, for all their support through the writing process: Kristín Linda Ragnarsdóttir; Guðlaug María Sveinbjörnsdóttir and her husband, Birkir Eyþór Ásgeirsson; Davíð Örn Sveinbjörnsson and his wife, Fjóla Dögg Halldórsdóttir, and their children, Salómon Blær, Elísa Björt, Annika Bára, and Mikael Björgvin. Thank you all for all your love and support!

We hope this book will promote safe and responsible work, securing that everyone gets back home safely.

Reykjavík, March 6th, 2022

Benjamín Ragnar Sveinbjörnsson
Sveinbjörn Gizurarson

CHAPTER 1

Introduction to safety

Safety mindset and safety cultures

One of the steps we can take to enhance our safety is to adapt a safety mindset. A safety mindset involves prioritizing safety first, practicing being aware of and proactively thinking about potential hazards in our surroundings, taking steps to minimize potential dangers that we notice, and preparing a game plan for what we would do in the event of an accident. Think about your experiments and work in advance with safety in mind, remaining vigilant of your surroundings, how well equipped and designed your facilities are from a safety perspective, and consider your coworkers' safety along with your own.

Equally important is cultivating a safety culture. It is important that you and your coworkers are comfortable in discussing any and all safety aspects of your work, and that you are at ease asking for advice regarding the safety of your experiments. Whenever doing a new experiment, it can be good to walk through the procedure with a coworker, noting the potential hazards you have noticed, and receive feedback on whether there are more hazards that you might have missed that you need to be prepared for and what steps would be best to take to minimize such dangers.

You should also make sure to communicate any issues that you notice to your supervisors. Unfortunately, accidents can happen because of factors that we did not expect to be a potential hazard. It is therefore important to report and discuss openly all accidents that do happen and think about ways to prevent similar accidents from taking place again. Let us all work together to get everyone safely home, each and every day.

Be familiar with the facilities

When you enter a laboratory, your first task should be getting familiar with the facilities. Where are the emergency exits, the sinks, emergency showers, eye wash stations, fire extinguishers, etc.? Are there sinks that are reserved

Handbook for Laboratory Safety
ISBN 978-0-323-99320-3
https://doi.org/10.1016/B978-0-323-99320-3.00002-1

for handwashing only or is the same sink also used for cleaning glassware? What personal protective equipment (PPE) are people required to wear in this workspace? If this is your first time working in a laboratory, be especially alert and observant, and make sure to get familiar with the new work environment and instruments there, especially those relating to safety, before you start working.

Never work alone in the laboratory

This is a crucial rule. You should never work alone in the laboratory! If a serious accident takes place, it is important that there are other people there who will be available to help immediately. If you are alone during a serious lab accident, there is no knowing when help might arrive, and odds are that it would arrive too late.

Since we should never work alone in the laboratory, it is also useful to be mindful of what others in the lab are doing. If you see something that looks unsafe, say something! Accidents in the laboratory can affect more people than just the lab worker doing the experiment that resulted in the accident. It is therefore important to say something if you see something, both for your own safety and your coworkers'. And remember that it is perfectly acceptable and should be encouraged to ask for advice if you have any questions regarding safety.

Food, beverages, and smelling in lab

It is strictly forbidden to eat or drink in the laboratory. There could be chemical traces anywhere in the lab and you should never take the chance of contaminating your food or drink by bringing it into the laboratory. This extends to all edibles including candy and/or chewing gum in your pockets (Fig. 1.1).

In general, you should never touch, taste, or smell chemicals that you are unfamiliar with. If you are familiar with them, you should probably not touch, taste, or smell the chemicals either, but in those cases, you would at least be more aware of the potential consequences. The smell will not tell you how dangerous a compound is. Many highly toxic compounds have no identified aroma. An example of that is carbon monoxide (CO), a colorless and odorless gas, which if exposed to, can cause loss of consciousness and even death (Fig. 1.2).

All work with harmful and toxic compounds should be carried out inside a well-ventilated cabinet or a fume hood to minimize the chance of

Figure 1.1 No drinks or food should be brought into the laboratory for consumption. *(Credit: Birkir Eyþór Ásgeirsson).*

Figure 1.2 You should never smell chemicals in lab unless you are absolutely sure they are safe. *(Credit: Birkir Eyþór Ásgeirsson).*

you unnecessarily breathing in fumes of hazardous compounds. Remember to lock all bottles as required[1], after use, and keep your workspace clean and organized so that you have a clear overview of everything you are working with, at every time point.

---

[1] Some bottles should be locked tightly, while others need to be able to allow pressure build-up to be released.

Smoking and alcohol

Smoking is also always forbidden in the laboratory and its surroundings. Due to the number of flammable solvents and chemicals, smoking would create an additional fire hazard and is therefore strictly prohibited inside the lab. It is also strictly forbidden to be intoxicated, no matter how slightly, while in the laboratory. Intoxication increases the risk of mistakes significantly, so it is never acceptable to work in lab intoxicated.

Personal state of mind

It is also important to be aware of our personal state of mind. Tiredness can have similar effect on us as alcohol, including slowing down our reaction times and reducing our concentration. You should therefore make a habit of getting proper sleep before working in lab and make sure you are feeling well awake, especially when doing something new or potentially hazardous in lab.

Our various states of mind can also create unnecessary hazards, whether it be carelessness, rashness, over confidence, or some other state of mind that alters our judgment and behavior in lab. It can be tricky for us to notice these behaviors in ourselves, but by knowing that these can affect our safety, we are more likely to think about it and do what we can to prevent it from creating unnecessary additional risks in our work.

> ### Examples from real life—Mindset
> Stay focused. If you are reading or browsing your phone, do that while standing still. More and more accidents are happening because people are trying to multitask. Walk, carrying samples, checking messages on the phone, all at the same time. Then they walk into a door or tip over a doorstep!

The importance of proper preparation

Your safety is heavily influenced by your knowledge of the procedures you are about to perform and how well you have prepared yourself before working on your project. Make it a habit to prepare properly and consider potential risks so you can prevent them. What could go wrong? How should I handle an accident or an emergency situation? How should I respond to a chemical spill? What if I get chemicals into my eyes? In the event of an accident, you may need to act immediately since every second may be critical. Therefore, it can be a good idea to walk through what you would do in case of an emergency. That can help getting it into your muscle memory which reduces the chances of panicking if the situation arises where you need to act quickly.

Proper preparation includes collecting information on the instruments, chemicals, bacteria, etc., that you will be working with. If you will be working with a *chemical*, you should know how it behaves. Is it flammable, corrosive, or maybe toxic? It is important to realize that only the most important information is found on the label, but you can find more detailed data in the safety data sheet (SDS) from the manufacturer. It is your responsibility to obtain the SDS and familiarize yourself with the information there *before* you start your work. Therefore, make sure to read *both* the labels and the SDS.

In general, make sure to handle all chemicals extremely respectfully. Always be alert when working with chemicals and have your mind on the project at hand. As research can often involve the synthesis of novel compounds, it is also important to know how to treat those substances. In case you are working with an unknown sample or compound, handle it as if it is the "most toxic compound" in existence until you know how it behaves. Never work with highly toxic compounds unless you have received the appropriate training.

You should also be familiar with your equipment and how to operate it. Always use appropriate personal protection such as a lab coat, gloves, safety glasses, and other appropriate safety equipment and precautions. It is extremely important that you wear the appropriate PPE for each experiment and putting on the relevant PPE should in fact be the first thing you do when you enter the lab.

PACE yourself

One approach to preparation with safety in mind is to use the following four-step approach, which the mnemonic "PACE yourself" might help you remember (Fig. 1.3):

(1) **Plan:** Write down the planned procedure in detail. Note down any risks that you notice right away and what needs to be done to minimize the risk. It can be good for workplaces to have a template for this type

Figure 1.3 The mnemonic PACE yourself may be useful to help you remember to Plan your experiment, Ask yourself questions about its safety, and Consult a coworker before you Execute all appropriate precautionary steps and the experiment itself.

of planning sheet that can then be placed close by the experiment while it is running, so people walking past can inform themselves about it. This is especially important for long running experiments where you might not always be close by in case something goes wrong.

(2) **Assess:** Assess the potential risks. This can often be done by asking yourself questions about the potential risks, so it can be useful to have a list of questions at hand for that. Such a list helps to remind us of possible hazards that we may not have thought about at first. Here follow a few examples of questions that might be useful, but it is recommended that you create your own list and add to it as you think of more hazards that you want to double check with yourself:

- How hazardous are the chemicals/biological samples/other material being used in the experiment (e.g., toxic, flammable, explosive, contagious)?
- Are the instruments and glassware in a good condition?
- Is the necessary safety equipment at hand in case something goes wrong?
- What side reactions might take place and what hazards would be associated with those? Are there any toxic byproducts or could there be a pressure build-up from gaseous byproducts in a closed system?
- Could any effects of the experiment reach far? Do people need to be notified of the experiment? Some chemicals have, e.g., a strong smell that might catch people off-guard and it would be good to notify them in advance if they might be affected by your experiment.

(3) **Consult:** It is always useful to consult with someone else regarding potential hazards that you might have missed. This works best when such a consultation is a normal part of the safety culture at your

institution. Be willing to ask for consultation and be willing to provide consultation when asked for it.

(4) **Execute precautionary steps:** Make sure to execute all appropriate precautionary and preventative steps necessary. Keep in mind that it is good when the precautionary steps include contingency plans. Then when everything is ready, go ahead and execute the experiment.

Keep it clean and organized

Never leave bags, clothes, or something else in the walkway, otherwise someone might walk into it and trip. You should also *never* run in the lab and it is recommended to avoid wet floors as they might be slippery.

Make sure your workspace is clean and organized, especially around balances, in fume hoods, and around any chemical storage. All bottles and packaging should be well labeled, clean and without strips, drops, or contamination. Clean your work area in order to prevent cross contamination to the next person working there or their experiment. If your chemical gets spilled, never pour it back into its bottle! Clean everything, including glassware, packaging, etc., and dispose of it according to the proper guidelines. Tables, floors, and fume hoods should be cleaned before you leave at the end of the day.

All work with flammable compounds, such as solvents, should be carried out inside properly ventilated fume hoods or cabinets. Flammable chemicals should *never* be heated with an open flame, but water/oil/sand baths, or other safe equipment can often be used instead if you need to heat them.

If you need to transport compounds from one place to another, make sure all bottles are closed and in a secondary container during transportation, and walk calmly while carrying the compounds. Never remove chemicals from the laboratories for your personal use or to make practical jokes. That may result in a warning and/or termination/expulsion.

At the end of the day

Make sure your workspace is clean and organized. If an employee, student, or a cleaning personnel places their hand on the table, will they get contaminated with a chemical, radioactive compound, or a microbe that may affect their health condition? Before you leave you should

- ✔ Turn off gas, water, and electricity.
- ✔ Turn off any instruments that do not need to be on.
- ✔ Clean your workspace.
- ✔ Make sure that all bottles are appropriately closed.
- ✔ Make sure that all chemicals are properly labeled and back where they should be stored.
- ✔ Close the fume hoods.
- ✔ Wash your hands.

Examples from real life—At the end of the day

The water alarm went on, one night. When the maintenance people arrived, there was water all over the floor, dripping to the next floor below. What happened was that a technician in the laboratory had turned on the water, but this day the water had been shut down due to maintenance. Unfortunately, this employee was not aware of that, but forgot to turn it off again.

Later that afternoon, a student had placed a cloth on the sink, and it fell into the sink. When the water maintenance was over, in the late afternoon, the cloth clogged the sink, filling it with water, and caused serious damage to the surroundings.

Always remember to leave your workplace in safe condition, at the end of the day.

CHAPTER 2

Personal protective equipment

Overview of personal protective equipment

One of the most important precautionary steps we can take in the lab is to use appropriate personal protective equipment (PPE). PPE should be the first thing you put on before or immediately upon entering the laboratory as it acts as a vital layer of defense in the event of an accident. The appropriate PPE can vary between fields, locations, and experimental settings, but is generally any protective wearables. It is important to keep in mind that the choice of PPE should not be solely dependent on the experiments you are doing, but you should also be mindful of experiments taking place in your vicinity. Reflective clothing, worn outside in the dark so that drivers can see us better, can be considered an example of general PPE that is worn because of potential hazards from others.

In the laboratory setting, there are three primary PPEs that should be worn most, if not all, of the time while in lab. These are safety glasses to guard your eyes, and a lab coat and gloves to guard your skin. In addition to these, you might need to wear other PPE depending on the setting. These could include masks, to guard your respiration, or a personal radiation detector, to monitor your radiation exposure. Make sure you always wear all of the appropriate PPE for the laboratory environment you are working in, as well as any additional PPE needed for the specific experiments you are working on. Here follows a more detailed discussion about the main types of PPEs worn in the lab.

Eye protection

Our eyes are one of the most valuable sense organs we have; therefore, it is extremely important that we make sure to keep our eyes as safe as possible! That means, we should *always* wear safety glasses/goggles while inside the lab. It is important to be aware that while conventional glasses may provide some protection, alas, they tend to cover only the area in front of our eyes

Handbook for Laboratory Safety
ISBN 978-0-323-99320-3
https://doi.org/10.1016/B978-0-323-99320-3.00006-9

and are thus not enough. Proper safety glasses provide additional protection on the sides as well (Fig. 2.1). Chemical splash safety *goggles* may provide the best protection as they reduce the chance of a major splash running down the face and into the eyes. If you use glasses for your daily life, you may need to wear safety goggles over those. Prescription safety glasses might also be available, and some institutions offer those for their employees, but they are not as easily available as safety goggles.

For employers, supervisors, and/or principal investigators, it is important to make sure that safety glasses are available for everyone working in the laboratory as well as for any visitors/guests that might be entering the laboratory, no matter how short term. Some laboratories have a small bin with extra safety glasses for short-time visitors to the lab.

The use of lenses is not recommended when working in a lab with chemicals. There is the danger of solvent vapors messing with the lens or getting trapped between the lens and your eyes causing irritation or damage. Some corrosive compounds could also accidentally splash into your eye(s) and the natural response to that is clamping our eyelids shut,

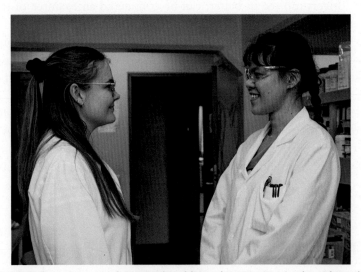

Figure 2.1 Safety glasses (right) provide additional protection on the sides, whereas regular glasses (left) do not provide that protection. Safety goggles (not pictured) provide even more coverage, reducing the chance of a splash running down the face into the eyes. Also note that long hair should be tied up in a tail (right) or kept under the lab coat while working in lab. If it is not kept out of the way (left), it can create unnecessary safety hazards. *(Credit: Birkir Eyþór Ásgeirsson.)*

making it really difficult to remove the contact lenses. All in all, it is generally better to just leave the contact lenses out of the lab.

The choice of safety glasses is important. For chemistry work, it is recommended to use chemical–resistant goggles, and when working with lasers, appropriate laser glasses are needed to protect against the wavelength that the laser emits. Laser accidents are always serious accidents. In general, it is important to choose the right protective eyewear depending on the work being done in the lab.

Proper eye protection[1]

A postdoctoral scholar was carrying out a scale-up of a reaction he had done before. The postdoc had recently changed the work-up procedure and tested it out on a small scale with no observable problems. However, during the work-up for the scaled-up reaction, there was a violent reaction that resulted in liquids splashing onto the researcher's face and lab coat. The postdoc was wearing safety glasses, gloves, and a flame-resistant lab coat with a long-sleeved shirt underneath.

After the splash, the researcher followed proper procedures, removing their lab coat and shirt, and went to the emergency shower where they rinsed off before heading to the eye wash station where they rinsed their face and eyes with the help of a graduate student.

The lab coat and shirt had done their job, minimizing chemical exposure to the body. The safety *glasses*, however, had not been enough. While they provided some protection, the chemicals were still able to run under the glasses, down the face and into both eyes. Chemical splash safety *goggles* would have provided more eye protection here. Thankfully, with the proper first reaction and subsequent treatment at the hospital, the researcher was cleared to return to work a few days after the incident.

[1] This case was extracted from Chance, B. S. Case study: Reaction scale-up leads to incident involving bromine and acetone. *Journal of Chemical Health & Safety*. 23 (1), 2—4, 2016.

Lab coat

The primary function of the lab coat is to protect your body and your clothes from contact with chemicals and other contact hazards in lab. Small chemical spills or splashes can often lead to holes in your clothes, so using a lab coat can actually save you from a bad accident and save you money on new clothes. In the event of a major chemical spill on yourself, the lab coat provides a removable barrier that you should be able to take off quickly.

The best lab coats have buttons that you don't need to take time to un-button in the event of an emergency, but that you can easily rip open quickly, sort of superman style. Not all of them are designed that way though, but it can be worth checking how easy it is to take your lab coat off quickly, so you are better prepared in the event of a real spill.

Since the lab coat is there to provide you with an outer layer protection against the chemicals in lab, be mindful of where and how you store it. For example, if you hang it up on top of another lab coat, the next time you wear it you might be putting on some chemicals from the outside of the other lab coat. Any cross-contamination of potentially hazardous chemicals is bad, especially when it is preventable. So, find a good place to store your lab coat in a way where the inside parts, that you will be in the most contact with, will stay safe and uncontaminated (Fig. 2.2).

Figure 2.2 Lab coats should be stored in a manner where the storage conditions do not result in cross-contamination to the inside of another lab coat. *(Credit: Benjamin Ragnar Sveinbjornsson.)*

Another benefit of lab coats is that they can help prevent unnecessary spread of contamination outside of lab. Just like we do not want their insides to get cross-contaminated, we also do not want their outsides to cross-contaminate things/areas outside of lab. The workplace often has "chemical free zones," the most important of which might be wherever people tend to eat. Remember to take off your lab coat before going to any of those common areas. In general, it is best to isolate the use of the lab coat to the areas and situations where they are needed. This can be especially important for biohazard labs.

You should also be mindful of the condition of your lab coat. Remember to wash it regularly, but make sure that it is not together with other clothes. If it gets contaminated with toxic compounds, wash it or destroy it. If it is starting to get old, tattered, and/or holey, it will not provide proper protection, so if that's the case, it is probably time to invest in a new lab coat.

There are several types of lab coats available, including splash-resistant, chemical-resistant, and flame-resistant lab coats. It is important to pick the appropriate type of lab coat that meets the needs of your experiments and lab environment. For example, if there is any chance of a fire, then you should not wear a lab coat made of synthetic fabrics as they can melt and stick to the skin when they burn. Many laboratories also have lab coats in different colors, depending on where they are used: in a regular laboratory, in an isotope laboratory, or in a lab where you work with biohazards. Therefore, make sure that you have the right lab coat(s) for your needs.

You should also make sure that your lab coat is a good fit from a size perspective. Lengthwise, it should reach at least down to your knees. For your arms, it should reach approximately to your wrists, so that it is not in the way when you are working, but so that it covers your whole arms down to your gloves. Some lab coats have fitted wristbands that help reduce the chance of a splash up the arm on the inside of your lab coat. Since we do want the lab coat to provide a full and proper protection, you should always wear it buttoned and never roll up the sleeves (Fig. 2.3).

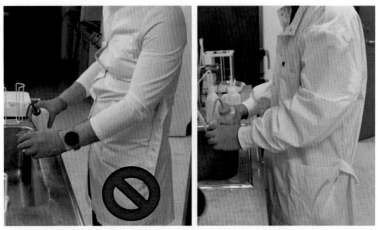

Figure 2.3 It is important to use a lab coat of an appropriate size and wear it properly so that it provides the best coverage and protection. *(Credit: Sveinbjörn Gizurarson.)*

Deadly fire at UCLA[2]

At the end of 2008, an accident that resulted in the death of a young lab worker at UCLA shook the chemistry world. This incident affected safety protocols and training at many higher education institutions in the following years and is still a prime example of the importance of proper safety training and the responsibility of the supervisors in providing such training.

A 23-year-old college graduate was only 3 months into her job in a chemistry lab when she was starting to extract 54 mL of *tert*-butyl lithium from a sealed container using a plastic syringe. This chemical can ignite instantly when exposed to air. During this process, the syringe came apart in her hands, spewing flaming chemicals, according to the accident report. She was not wearing a flame-resistant lab coat and the flash fire set her clothing ablaze and spread serious burns over 43% of her body, resulting in her eventual death 18 days later. There were several more safety protocols that might have helped but were neglected, a lot of which were traced to lack of proper safety training. This case serves as a tragic reminder of the importance of providing/receiving proper safety training for all lab workers, as well as the importance of wearing the appropriate PPE, as a removable flame-resistant lab coat might have reduced the burn injuries.

[2] Extracted from news reports including a report from Toronto Star, *A young lab worker, a professor and a deadly accident* by Kate Allen, published March 30th, 2014 and accessed October 27th, 2021.

Figure 2.4 Always wear closed toe shoes and long pants in lab. *(Credit: Sveinbjörn Gizurarson.)*

Shoes and long pants

Since the lab coat does not provide protection all the way down to your feet, it is important to wear closed toe shoes when working in lab to protect your feet. Do not wear open sandals, high heeled shoes, or other types of open shoes inside the laboratory. If you spill something, it will most likely drop down on your toes or feet, so use appropriate shoes. The above also applies to your legs. Therefore, use long pants/trousers and not shorts.

Some shoe materials may induce static electricity, which could cause damage to sensitive instruments, or even cause chemicals to burst in flames, so also keep that in mind when choosing what shoes to wear in lab (Fig. 2.4).

Gloves

Your hands are precious so use gloves to protect them! Their importance cannot be underestimated. Still, gloves cannot protect against all chemicals, which is why it is important to remain vigilant and also be familiar with what type of gloves work best for the type of work that you will be doing. Would a latex or a nitrile glove work better? Would disposable gloves be sufficient, or would a thicker, more durable glove be more suitable? The chemicals that you will be working with will have different permeabilities through different types of gloves, so it is worth checking it out in advance. While gloves serve as an important PPE, it is also important to remain vigilant and realize that some chemicals can even go through certain types

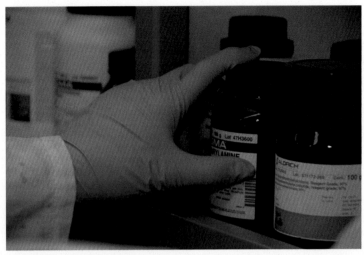

Figure 2.5 Make sure you wear appropriate gloves for the material you will be working with. *(Credit: Sveinbjörn Gizurarson.)*

of gloves without leaving a trace and result in real harm (see *Deadly drops of dimethylmercury* in Chapter 4) (Fig. 2.5).

In general, you should always use gloves when working with corrosives, toxic compounds, and biologicals such as blood samples or microbes. It is also important to be mindful of what surfaces you touch with the gloves on. *Do not* open any doors with your gloves on, nor operate a water fountain, nor use your gloves on any other common area that people touch without gloves. That could result in cross-contamination that could become harmful for the next person. It is safer and simpler to throw used disposable gloves away in the trash and put new ones on.

With the advent of smartphones, it has also become increasingly common to see people in lab operate their phones with their gloves on. This is not a wise thing to do! The trace chemicals that may be on your gloves could then journey to your phone and later, when you answer a call and put the phone up against your face, they might end up there. Just take off your gloves before using the phone and think about the safety aspects of *everything* you do in lab.

Jewelry should not be worn when working with biological materials or sterile samples. Microbes are well protected between the skin and the jewelry and even after a thorough hand wash, it is difficult to eliminate all microbes. You should also remove all jewelry if you are working with compounds that may react with gold or mercury.

In order to help you become accustomed with the different types of materials in your gloves and their resistance for the compounds you are working with, here are some guidelines:

- ✔ *Temperature-tolerant gloves*: These gloves should tolerate very high or very low temperatures, such as cryogenic fluids (liquid N_2). Examples are Kevlar gloves, aluminum-coated gloves providing very good isolation against extreme temperatures and Zetex gloves, synthetic gloves resistant to high or low temperature as well as certain corrosives.

- ✔ *Chemically resistant gloves* are gloves made of different types of polymers such as latex, nitrile, butyl rubber, or neoprene:

  o *Butyl gloves* are resistant to acids such as nitric, sulfuric, and fluoric acid as well as highly oxidizing compounds. They are also resistant to gases, fumes, and numerous other compounds. They do not deform in cold temperature.

  o *Latex gloves* provide good protection for aqueous solutions, dilute acids and bases, salts, and ketones. They are very comfortable, but all latex gloves are able to induce irritation, rash, or allergies. Some high-quality latex gloves release less particles, resulting in lower risk of allergy. It is not recommended to use latex gloves containing powder. Such gloves may cause even more problems if you do not tolerate latex, because the particles get mixed with the powder that gets easily distributed over a larger area, causing irritation in different mucosal surfaces. Additionally, the powder may be carried with you on your clothes or contaminate your work. It is recommended to use gloves with less than <100 µg/g extractable latex particles.

  o *Neoprene gloves* give excellent protection against organic solvents, oils, and organic acids and bases. These gloves are elastic and durable.

  o *Nitrile gloves* give excellent protection against chlorinated compounds such as trichloroethylene and perchloroethylene. These gloves are good for fine work. They are resistant, durable, and elastic.

  o *Vinyl gloves* should generally not be used for work with hazardous chemicals or microorganisms as they provide comparatively little protection.

Chemically resistant gloves are often available both as thinner gloves, considered to be disposable, or in a thicker version, intended to be reusable. For the disposable gloves, it is often especially important to change them in the event of a spill, to minimize the risk of the chemical spill having enough time to permeate the glove. The thicker, reusable gloves often have a cotton flock liner inside the glove and can be rinsed to a certain extent after a spill, depending on the spill. These gloves are often used for washing and recovering glassware from acid or base baths. It is especially important to be aware of any potential leakage in the gloves. Often these gloves develop tiny rips around the fingers and hands that allow for leakage, so that they stop providing the proper protection. If that occurs, the glove should be taken out of use so that it does not provide a false sense of security for the next, unsuspecting user.

Here below is a table with information about the most common glove types and their resistance to certain chemicals. The glove types are generally rated based on how long it takes for the chemicals to permeate the gloves. Although this table can serve as a useful guide, do not rely solely on it. Make sure to double-check the information given by the manufacturer, since important aspects, such as glove thickness, can vary between manufacturers and affects the resistance (Table 2.1).

Table 2.1 Chemical resistance of a few common glove materials. *A*, acceptable; *G*, good; *P*, poor; *VG*, very good. Empty entrances indicate that the gloves have not been tested yet or the information is unknown.

Compounds	Neoprene	Latex	Butyl	Nitril	Vinyl (PVC)
Acetaldehyde	VG	G	VG	G	P
Acetic acid	VG	VG	VG	VG	A
Acetone	G	VG	VG	P	P
Acetonitrile		A	G	P	P
Ammonium hydroxide	VG	VG	VG	VG	VG
Amyl acetate	A	P	A	P	P
Aniline	G	A	A	P	A
Battery acid		G		VG	VG
Benzaldehyde	A	A	G	G	P
Benzene	P	P	P	A	P
Boric acid		G	G	G	
Butane		P		VG	P
Butyl acetate	G	A	A	P	P
Butyl alcohol	VG	VG	VG	VG	G
Cadmium oxide	G			G	
Carbon disulfide	A	A	A	A	P
Carbon tetrachloride	A	P	P	G	P
Castor oil	A	P	A	VG	VG
Chloroacetic acid	G		G		
Chlorobenzene	A	P	A	P	P
Chloroform	G	P	P	A	P

Table 2.1 Chemical resistance of a few common glove materials. *A*, acceptable; *G*, good; *P*, poor; *VG*, very good. Empty entrances indicate that the gloves have not been tested yet or the information is unknown.—cont'd

Compounds	Neoprene	Latex	Butyl	Nitril	Vinyl (PVC)
Chlorophane	A	P	A	A	P
Chromic acid (50%)	A	P	A	A	G
Citric acid (10%)	VG	VG	VG	VG	VG
Cresol	G	P		G	A
Cyclohexanol	G	A	G	VG	G
Dibutyl phthalate	G	P	G	G	G
Diesel fuel	G	P	P	VG	
Diisobutyl ketone	P	A	G	P	P
Dimethylformamide	A	A	G	P	P
Dimethyl sulfoxide (DMSO)			G	G	P
Dioctyl phthalate	G	P	A	VG	P
Dioxane	VG	G	G	G	P
Epoxy resins (dry)	VG	VG	VG	VG	
Ethanol (alcohol)	VG	VG	VG	VG	G
Ether (diethyl ether)	VG	G	VG	G	P
Ethyl acetate	G	A	G	A	P
Ethyl alcohol (ethanol)	VG	VG	VG	VG	G
Ethylene dichloride	A	P	A	P	P
Ethylene glycol	VG	VG	VG	VG	VG
Ethyl ether	VG	G	VG	G	P
Ferrous sulfate	G	G		G	
Formaldehyde	VG	VG	VG	VG	VG
Formamide	G		G		
Formic acid	VG	VG	VG	VG	VG
Freon	G	P	A	G	P
Furfural	G	G	G	G	P
Gasoline (leaded or unleaded)	G	P	A	VG	P
Glutaraldehyde	G	G	G	G	
Glycerin	VG	VG	VG	VG	VG
Hexane	A	P	P	G	P
Hydrazine (65%)	A	G	G	G	VG
Hydrochloric acid	VG	A	G	G	VG
Hydrofluoric acid (48%)	VG	G	G	G	VG
Hydrogen peroxide (30%)	G	G	G	G	VG
Hydroquinone	G	G	G	A	VG
Isoamyl alcohol	G		G	G	
Isobutyl alcohol	G	VG	G	G	A
Isooctane	A	P	P	VG	P
Isopropyl alcohol	VG	VG	VG	VG	G

Continued

Table 2.1 Chemical resistance of a few common glove materials. *A*, acceptable; *G*, good; *P*, poor; *VG*, very good. Empty entrances indicate that the gloves have not been tested yet or the information is unknown.—cont'd

Compounds	Neoprene	Latex	Butyl	Nitril	Vinyl (PVC)
Kerosene	VG	A	A	VG	A
Ketones	G	VG	VG	P	
Lacquer thinners	G	A	A	P	
Lactic acid (85%)	VG	VG	VG	VG	VG
Lauric acid (36%)	VG	A	VG	VG	A
Linoleic acid	VG	P	A	G	G
Linseed oil	VG	P	A	VG	VG
Lubricating oil	G	G		G	
Maleic acid	VG	VG	VG	VG	G
Methyl acetate		P		P	P
Methyl alcohol (methanol)	VG	VG	VG	VG	G
Methyl amine	A	A	G	G	VG
Methylene bromide	G	A	G	A	P
Methylene chloride	P	P	P	P	P
Methyl ethanolamine	G		G		
Methyl ethyl ketone	G	G	VG	P	P
Methyl isobutyl ketone	A	A	VG	P	P
Methyl methacrylate	G	G	VG	A	
Mineral oil		P		VG	A
Monoethanolamine	VG	G	VG	VG	VG
Morpholine	VG	VG	VG	VG	P
Naphthalene	G	A	A	G	P
Naphthas—aliphatic	VG	A	A	VG	P
Naphthas—aromatic	G	P	P	G	P
Nitric acid	P	P	P	P	P
Nitric acid 30%—70%	G	A	A	A	A
Nitrobenzene		P	G	P	P
Nitromethane	A	P	A	A	P
Nitropropane	A	P	A	A	P
Octyl alcohol	VG	VG	VG	VG	A
Oleic acid	VG	A	G	VG	A
Oxalic acid	VG	VG	VG	VG	
Palmitic acid	VG	VG	VG	VG	G
Perchloric acid (60%)	VG	A	G	G	VG
Perchloroethylene	A	P	P	G	P
Perfluoropentanoic acid	G	G	G	G	
Petroleum distillates	G	P	P	VG	
Phenol	VG	A	G	A	G
Phosphoric acid	VG	G	VG	VG	G
Picric acid	G	G		G	VG
Potassium hydroxide	VG	VG	VG	VG	VG

Table 2.1 Chemical resistance of a few common glove materials. *A*, acceptable; *G*, good; *P*, poor; *VG*, very good. Empty entrances indicate that the gloves have not been tested yet or the information is unknown.—cont'd

Compounds	Neoprene	Latex	Butyl	Nitril	Vinyl (PVC)
Propyl acetate	G	A	G	A	P
Propyl alcohol	VG	VG	VG	VG	A
Sodium cyanide (solid)	G	G		G	
Sodium fluoride	G	G		G	
Sodium hydroxide	VG	VG	VG	VG	G
Sodium hypochlorite	G	G	G	G	
Sodium silicate		G	G		
Styrene	P	P	P	A	P
Sulfuryl chloride			G	G	
Sulfuric acid	G	G	G	G	G
Tannic acid	VG	VG	VG	VG	VG
Tetrahydrofuran	P	A	A	A	P
Toluene	A	P	P	A	P
Toluene diisocyanate	A	G	G	A	P
Trichloroethylene	A	A	P	G	P
Triethanolamine	VG	P	A	VG	VG
Turpentine	G	A	A	VG	P
Vegetable oil		P		VG	A
Xylene	P	P	P	A	P

Respiratory protection and masks

It is important to minimize the risk of breathing in dust, gas, fumes, or aerosol in your lab work. Working in properly ventilated fume hoods can help in many situations, but sometimes it is also necessary to use appropriate respiratory protection. Masks may work in most situations when working with compounds that generate dust or particles, such as silica gel, and should therefore be used in those situations. Gas masks will protect you from many harmful gases, fumes, or fine dust. In the most extreme cases you may need a tight gas mask connected to oxygen. Some of those only cover the mouth and nose, but others cover the whole face (Fig. 2.6).

How do you select the right respiratory protection? Several factors are important before making that selection. These factors include resistance, maximal exposure limit (MEL), and threshold level (TLV). MEL is defined as the maximal exposure time limit before the mask will start to leak. TLV is the maximal concentration allowed in the atmosphere for the mask to still offer protection. It is also important to consider the types of fumes, gases, or dust that you are protecting yourself against. Are they corrosive, solvents,

Figure 2.6 Some research may require the use of a mask as part of your PPE. *(Credit: Sveinbjörn Gizurarson.)*

toxic, radioactive, carcinogenic, etc., or does it contain compounds such as asbestos? Is the gas life threatening? No single mask can protect against all these substances in one unit.

Most masks protect both your nose and mouth. Respirator masks are often classified into three categories, N, R, or P. N indicates that the mask is "not resistant to oil," R stands for "resistant to oil," and P stands for "oil proof." These codes are followed by a number such as 95, 99, or 100 depending on their resistance. For example: N-95 filters at least 95% of airborne particles but is not resistant to oil, while P-100 is 99.97% resistant to most fumes, including oil, and it also contains a so-called HEPA (high-efficiency particulate air) filter. Another classification for the efficiency of masks is bacterial filtration efficiency (BFE) and viral filtration efficiency (VFE). If you are protecting yourself against bacteria or viruses, you should note the BFE and/or VFE percentages for the mask types you are using.

Gas masks that allow you to attach different filters and multiple filters give the most effective respiratory protection. These filters are color-coded, where each color indicates their resistance (Table 2.2).

It is important to attach the masks properly and note the time you open the mask so you can keep its lifetime in mind. Do not exceed the MEL.

Table 2.2 Color-coding of filters for gas masks. As the color-coding might vary slightly depending on world region, make sure to verify whether this table applies to your location.

Color	Type of chemicals
White	Acidic fumes
White with a green line	Fumes that may contain cyanide
White with a yellow line	Chlorine gas
Black	Fumes from organic solvents
Green	Fumes from ammonia
Green with a white line	Fumes containing acid and ammonia
Blue	Carbon monoxide
Yellow	Fumes containing acid and organic solvents
Yellow with a blue line	Hydrocyanic acid gas and chloropicrin vapor
Brown	Fumes containing acid, organic solvents, and ammonia
Magenta	Particles containing radioactive materials (except tritium and noble gases)
Red with gray line	Fumes containing acid, organic solvents, ammonia, chlorine, and carbon monoxide
Purple	Any particulates—P100.[a]

[a] Work with mold and asbestos requires P100 rating. The HEPA filters used here may however not be as good in blocking certain fumes and gasses as some of the others mentioned here above.

Each time a new filter is used, make sure it is truly new and intact, and use it according to guidelines. Each filter has its expiration date. Used masks and filters should be disposed of.

Hair

It is recommended that long hair be tied up in a tail or kept under the lab coat while working in lab (see Fig. 2.1). Long hair may get stuck in rotating instruments, absorb chemicals, or catch on fire. If you are working with biological samples, protect your hair so it will not absorb anything from what you are working with. This is also important if your work may produce aerosol or fumes. Clothes that are hanging out from your lab coat may also get contaminated by absorbing unwanted biological matters or chemicals. We even see that when people with, e.g., pollen allergies get inside, their symptoms remain until they have changed clothes and washed their hair and removed most of the allergens from their environment.

Lead apron and shields

If you are working with radioactive materials or X-rays, it is important to limit the rays you will be exposed to. An appropriate shield and/or lead aprons are among the PPE that you can use to limit radiation exposure. The external and internal sex organs must be protected. Radioactive substances must also be stored appropriately, so as to prevent radiation from extending beyond the container. These are often stored in a container with a Plexiglas shield covering or in some cases lead (see more in Chapter 6).

Exercise

You have been asked to go to the warehouse and get 2 L of acetone from a drum. Select the appropriate PPE?

Exercise

How many safety violations can you see in the following figure? (Fig. 2.7)

Figure 2.7 There are numerous safety violations in this figure. Can you find them all? *(Credit: Birkir Eyþór Ásgeirsson.)*

CHAPTER 3

The laboratory

Hazards in the laboratory

While working in the laboratory, we are often surrounded by more hazards than in most other workplaces, especially when we are working with the unknown. A lot of chemicals are flammable, even explosive and if you are not using safe techniques and showing the utmost care a lot of things can go wrong, e.g., generation of poisonous gas/fumes. Biological hazards can spread over the workplace and you or someone close by might get infected by the biological samples being used.

You can often find chemicals from most or even all hazard classifications in laboratories at universities, research institutions, and in the industry. Some of the chemicals can be carcinogens or teratogens. Some laboratories work on the synthesis of novel compounds or work with previously un-known organisms, where no one knows what properties the chemicals and/or microbes have. It is not just enough that the facilities and the equipment fulfill all safety measures but it is also important that everyone working in the lab follows the appropriate guidelines and is aware of the potential hazards from the chemicals, equipment, and anything else present in the lab.

Safety is the most important thing in the laboratory. New staff and students in labs are in the most danger. This new workplace is alien, and it can be difficult to realize where the dangers can lie. Where are the most hazardous chemicals stored? Even experienced scientists are in danger if they are careless or distracted. It is therefore important to get well acquainted with the facilities and familiarize oneself properly with the tasks that await you. Simple and relatively safe chemicals can become hazardous if used incorrectly or in a careless manner. In this chapter, we will cover some protocols and make note of some things to avoid. Awareness of the hazards increases your own safety as well as the safety of your coworkers. Remember to always wear the appropriate PPE when working in the lab (Chapter 2). Also remember to never work alone in the lab.

Handbook for Laboratory Safety
ISBN 978-0-323-99320-3
https://doi.org/10.1016/B978-0-323-99320-3.00007-0

Thankfully, safety training is considered a prerequisite in most places before new staff and students are allowed to enter the laboratory. However, accidents can still happen and therefore it is important to stay focused, be careful in everything you do, and not get distracted. Broken glassware is responsible for most of the cuts in laboratories. Unfortunately, chemical accidents are way too common as well. Accidents unrelated to the laboratory setting per se, such as falling because of slippery floors, cracks, or unevenness in the floors, also happen regularly in laboratories, so be careful. It is of concern, when statistics are studied, how many accidents take place involving the eyes, where the use of safety glasses or safety goggles would have had a significant protective value. With better instrumentations, new problems have also arisen such as staying in the same position for prolonged periods of time or repetitive movements, e.g., when working on sample preparation.

Safety features in lab

In most university labs, research institutions, as well as in the industry, you should be able to see what dangers lie within *before* you enter. This is often done by having a simple sign up that has information on the equipment inside, as well as what types of chemical or biological hazards are in the lab (see Fig. 3.1). All laboratories have their own distinctive feature. Some are designed specifically for work with microorganisms, others for chemical synthesis, or work with lasers, etc. This means that the standards for them can be quite different and it is your responsibility to familiarize yourself with them, as well as the employer's responsibility to provide proper training and information needed, and make sure that the lab fulfills the required safety standards. All safety features and protections should be set up with regards to the work the lab space is intended for.

The first thing you should note when entering a new laboratory is where the main safety equipment is located. Where are the sinks, and is there a designated sink for handwashing? Where is the emergency shower and eye wash station? Make sure you know how to operate these so that you are prepared before an accident might happen, rather than having to figure it out during an accident. It is also important that these safety features be checked regularly. If the shower/eye wash station are not used for a long period of time, the first batch of water that comes through the pipes might be rusty and it is not nice to get dirty water all over yourself during an emergency. Depending on the type of shower/eye wash station, it might

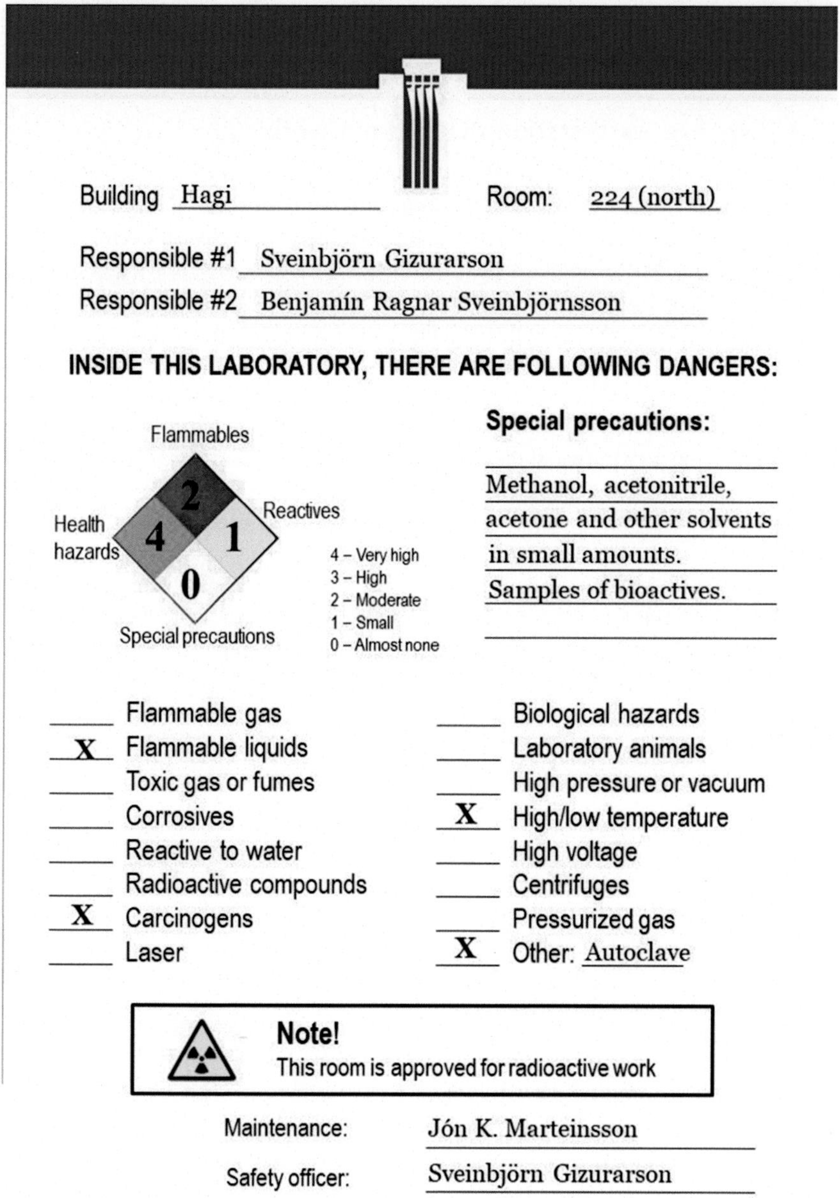

Building Hagi Room: 224 (north)

Responsible #1 Sveinbjörn Gizurarson

Responsible #2 Benjamín Ragnar Sveinbjörnsson

INSIDE THIS LABORATORY, THERE ARE FOLLOWING DANGERS:

Flammables 2

Health hazards 4

Reactives 1

Special precautions 0

4 – Very high
3 – High
2 – Moderate
1 – Small
0 – Almost none

Special precautions:

Methanol, acetonitrile, acetone and other solvents in small amounts.
Samples of bioactives.

	Flammable gas		Biological hazards
X	Flammable liquids		Laboratory animals
	Toxic gas or fumes		High pressure or vacuum
	Corrosives	X	High/low temperature
	Reactive to water		High voltage
	Radioactive compounds		Centrifuges
X	Carcinogens		Pressurized gas
	Laser	X	Other: Autoclave

Note!
This room is approved for radioactive work

Maintenance: Jón K. Marteinsson

Safety officer: Sveinbjörn Gizurarson

Figure 3.1 An example of a safety label, to be placed outside the laboratory, so in case of emergency you can quickly identify the risks present before entering. *(Credit: Sveinbjörn Gizurarson.)*

even become so rusty that it does not work properly when needed, as the case study below describes.

A rusty eye wash station that did not work[1]

One evening, an assistant professor of chemistry was working in lab with a coworker when he felt like something might have gotten into his eye. He was wearing safety glasses, but not safety goggles that could have prevented this incident. Unsure of whether it was just dust or something else that he got into the eye, he decided to wash his eyes to be safe.

The eye wash stations in this building were connected to the sink faucets in such a way that you had to push a small metal rod on the faucet to change the water pathway from the faucet itself to the eye washers. Unfortunately, they had not been tested regularly and neither the researcher nor his coworker could push it all the way to switch the water stream to the eye washers. They went to the next eye wash station, but it had the same problem. Thankfully the sink was big enough that his coworker could help him wash his eyes using the regular faucet and no damage was done to the eyes. Afterward they tried again to see if the eye wash stations worked and eventually got it to work but needed a wrench to hammer the rod through. As a follow-up, the eye wash stations were tested at a more regular interval and have since been changed for a system that does not rust as easily.

[1] This case is a personal experience of the authors who were the two researchers mentioned in the story.

You should also make sure that you know the locations of other safety equipment, such as fire extinguishers, fire blankets, spill kits, etc., before you start working in the lab.

All laboratories should also have at least two potential escape paths, and it is important to know where the nearest exits are in case of an emergency that requires quick evacuation. The arrangement in the lab, and the building, should be such that the escape path is always clear of cables or any items that might hinder or otherwise slow one's escape outside. Thresholds can also create an unnecessary tripping hazard, so if possible, they should not be present in laboratories.

Primary work environment

The working environment should be designed with the tasks in mind that will be conducted there. The floor, and ideally the walls as well, should have a chemically resistant surface that is easy to clean. The tabletops should be made from material that can withstand a short-term contact with various chemicals, including acids and bases. No tabletop is expected to leave no trace from contact with any chemical, e.g., if an acid has burned its way through the surface. It is preferrable that the surfaces be nongloss rather than shiny, as it may otherwise have more effect on the reading of results, e.g., when distinguishing between colors. Most laboratories need to allow for both sitting and standing working conditions. Chairs/stools should be adjustable and with clothing that is easy to clean.

Fume hoods and air conditioning are important in laboratories to remove all vapors and gases that can form. All work with volatile chemicals should be done inside a fume hood and work with biological hazards should be done in biological safety cabinets (BSCs). The air conditioning of the fume hood needs to be powerful enough to prevent the researchers from inhaling vapor from the chemicals being used, and also prevent the vapor density of the solvents from becoming so high that they might start burning spontaneously. The minimum air flow standards for fume hoods vary between the fume hood sizes, but the fume hoods and BSCs should be checked regularly to see if they meet the industry and manufacturer's standards. It is also important to keep in mind that the fume hoods and BSCs will not do their job if they are not used properly. Try to keep the hoods closed as much as possible and if you need to open them past the suggested maximal operating height, do it briefly and at a time when the contents inside the hood do not pose a danger to yourself or others.

In general, it is important to keep your workspace clean and organized. A well-organized workspace can both enhance the safety of your work environment and help you be more efficient in your work as you do not need to spend as much time searching for things. It is also important that the work environment be well lit, so it is important to make sure that the lighting fulfill necessary requirements.

Biological storage

All cabinets, refrigerators, and freezers that are used to store biological samples should be well labeled as such and disinfected regularly. They should always be kept in the appropriate biosafety laboratories (BSL-1, BSL-2, BSL-3, or BSL-4) to limit access to the samples to only those permitted to work with them.

Chemical storage

An immense amount of chemicals is used in laboratories, especially at research institutes and universities. Often there are special chemical storage rooms that are set up in the research facilities, to store the chemicals in a secure manner. Within the laboratory itself, chemicals tend to be stored on the shelves or in specially designed ventilated cupboards. When chemicals are stored on a shelf, it is important that the shelves are not open so that the chemicals do not fall off easily. In general, all chemicals should be stored in a secondary container or area that can act as a secondary container (e.g., the cupboards). The storage should be thought out in a way to minimize the damage from a potential chemical spill in the event of an accident, such as in case of an earthquake. It should also provide an easy access to read on the labels and all containers should have the correct warning labels.

When organizing the chemical storage, make sure to follow the guidelines that come with the product, usually written on the label, e.g., the hazard and precautionary statements, about storage conditions. What temperature should the chemical be stored at? Does it need to be stored under inert atmosphere?

Most labs have flammable safety cabinets, in which all flammable solvents and chemicals should be stored, including most organic solvents. Flammable chemicals should also never be in the proximity of instruments that can give a spark or have hot wires.

Another important consideration is compatibility with neighboring chemicals. For example, we want to store acids separate from bases since they can react violently with each other in the event of an accident, or in a manner that produces toxic products. Take those aspects into consideration when organizing your storage. Below are some basic guidelines that may be of use but given that many chemicals might fall into more than just one hazard category, it is still important to review the safety data sheets for further guidance on storage requirements.

- *Acids*: Should be stored in special acid cabinets and need to be stored separately from bases. The different types of acids, such as organic acids and oxidizing acids, should also be segregated. Acids should also be stored away from active metals (lithium, sodium, etc.) and chemicals that could generate toxic gases (e.g., sodium cyanide).
- *Bases (alkalines)*: Should be stored away from acids but can be stored in similar cabinets.
- *Flammables*: Should be stored in appropriate flammable storage cabinets and should always be kept away from potential ignition sources. They should be stored separately from oxidizers and oxidizing acids.

- *Oxidizers*: Need to be stored separately from organics as they can react violently with them. Keep away from combustible materials, reducing agents, organic solvents, and other flammables. It is also beneficial to store them in cool, dry places.
- *Pyrophoric substances*: These compounds can spontaneously ignite in air so they need to be stored in an air and water free environment and away from flammables. Cool and dry places are beneficial for these compounds.
- *Water reactive chemicals*: These compounds need to be stored in a water-free environment as they react violently with water. This can include some of the alkali metals and they often come in containers where the metal is immersed in oil to keep it safer. Because of the danger of violent reactions, these should be stored away from flammables in a cool, dry place.

The chart in Fig. 3.2 also provides some rough guidelines about chemical compatibility, but remember to double-check the safety data sheets, keeping in mind that some chemicals can fall in more than one category.

Some chemicals need to be stored at lower temperatures, either in a refrigerator or a freezer, either $-20°C$ $(-4°F)$ or even at $-80°C$ $(-112°F)$. When storing chemicals in a refrigerator or a freezer, it is important to make sure that the cooling unit fulfills the necessary safety standards for the chemicals being stored there. If those include potentially explosive compounds, then a refrigerator with explosion proof interior should be used, or in the very least, a spark-free unit.

A refrigerator fire[2]

In 2016, a fire erupted in one of the teaching labs at the University of Iceland. Thankfully it happened at a time when no one was working in the lab, but it resulted in serious damage to the laboratory and instruments stored there. Upon examination, it was believed that the fire originated in a refrigerator. In 1998, another fire had erupted from a refrigerator in a neighboring lab. Since then, the University has switched out most of the refrigerators used for chemicals, for explosion-proof refrigerator units.

[2] This case was extracted from an Icelandic news story from RÚV about this fire: Sigurðarson, G. *Talið að kviknað hafi í út frá ísskáp [Fire suspected to have ignited from a refrigerator]*, published October 25th, 2016 and last accessed October 31st, 2021. https://www.ruv.is/frett/talid-ad-kviknad-hafi-i-ut-fra-isskap.

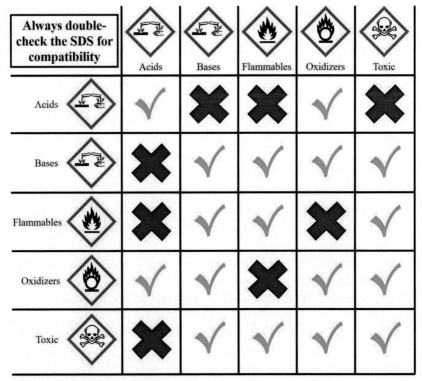

Figure 3.2 A chart providing rough guidelines about chemical compatibility. Make sure to store incompatible chemicals separate from each other. *(All hazard pictograms were obtained from the Globally Harmonized System of Classification and Labelling of Chemicals (GHS), 9th ed. and used with the permission of the UN.)*

After use, remember to put all chemicals back in their proper storage space in the laboratory and make sure that the working area, storage area, and the containers are adequately clean.

Gas storage and gas use in lab

Gas pipes for flammable gases should only be in areas where it is permissible to work with open fire. The use of fire in the laboratory is always contingent on nothing flammable being in the flame's proximity. If there is a need to heat up solutions of flammable solvents, that should be done using hotplates (or other adequate heating substitute) and inside a fume hood. It is important that in labs where people work with potentially hazardous gases, that there also be the relevant gas detectors, preferably connected to the building's security system.

Many labs use gas stored in high pressure gas cylinders, e.g., argon or nitrogen for work under inert atmosphere. These gas cylinders should be fastened properly, either in an appropriate gas cabinet or to the wall (Fig. 3.3). In areas where there is a high risk of earthquakes, it is advisable to have the gas cylinders fastened in two locations. If these high-pressure gas cylinders are not secured, there is the danger of them falling over, getting damaged and rapidly depressurizing. In one of their episodes, the Mythbusters tested whether a rapidly depressurizing gas cylinder could break through a wall. In their experiment, the "air cylinder rocket" flew through a concrete wall and made a significant dent in the next wall, about 2 meters (six feet) behind it as well.

Glassware and general storage

A lot of glassware, consumables, and tools are also stored in laboratories and should be stored in an organized manner that minimizes the risk of accidents if items or glassware fall out of their shelves, for example. Be extra

Figure 3.3 All high-pressure cylinders should be securely fastened to prevent them from falling over. The gas cylinder on the left is not fastened at all to the wall, while the cylinder on the right is fastened to the wall using a special strap. *(Credit: Benjamín Ragnar Sveinbjörnsson.)*

mindful of earthquake hazards if you are working in an area where earthquakes are common, as you do not want glassware to fall all over you. All storage shelves also need to be well fastened and should not be over-loaded with heavy items that might affect whether or not the shelf can adequately support the weight.

Transportation in or between laboratories

All transportation of chemicals to and from the laboratory should be done in as safe of a way as possible. The chemicals should always be transported in a secondary container and it is good to use trays as well whenever possible. It is important to take special care to use a secondary container for the transportation of strong acids and/or bases to ensure that they do not fall on the floor and cause damage.

The same applies to transportation of gas cylinders and cryogenic liquids. If an elevator is used for the transportation of a cryogenic liquid, it is strongly recommended that no one accompany it in the elevator. When the liquid evaporates, it can push the air (including oxygen in the air) away, which creates a hazardous situation.

Electrical safety

Almost all of the instruments we use are electrically powered and, in light of the chemical and/or biological hazards in lab, it can be easy to forget that we also need to be mindful of electrical safety.

When it comes to electrical safety, it is important to know the location of electrical panels so that you can quickly shut off the electricity in the event of an emergency. There should always be clear access to these panels so make sure that the surrounding area is not cluttered. In the event of a power loss, these panels can also be a good place to check to see if it is just an outage within the lab or if the power outage might be a more significant one in the neighborhood. It is always good to have a plan ready for various circumstances, including a power outage. In a power outage, the fume hoods stop providing proper ventilation so you should not use them at those time, and you also need to be careful about potential vapors from the chemicals present inside the fume hoods. Temperature-sensitive chemicals stored in refrigerators and freezers can also create hazards as they might warm up too much so think about what you would need to do to limit any dangers.

All equipment that depends on electrical power supply should be checked regularly. Also, a common electrical safety issue, not only in labs but also in many homes, relates to power cords. Power cords need to be checked to ensure that they do not have exposed wiring and are placed in a neat secure manner, so they do not come in contact with water or chemicals. They should be placed so they do not unnecessarily clutter your workspace or create a tripping hazard when on the floor. Nor should they be left in a way where people can easily get caught up in them, e.g., if they are dangling from a hood which might result in someone accidentally unplugging the equipment and/or causing it to fall over. The cords should be kept away from hot surfaces as we do not want the insulation to melt. Extension cords should be avoided and never be in permanent use in the lab. Be careful of overloading circuits by plugging too many high current drawing devices into the same outlet (Fig. 3.4).

Figure 3.4 Cluttered power strips in a classroom with cords all over the floor, creating an unnecessary tripping hazard. *(Credit: Sveinbjörn Gizurarson.)*

Make sure to keep flammable materials away from electrical equipment as they can serve as a source of ignition. This is especially important if the electrical equipment produces any sparks. Being mindful of electrical safety is important as electrical fires are serious on their own, but in a lab they can easily lead to a much more serious accident.

A few useful laboratory instruments

Hot plates can and have led to fires in laboratories. Those cases tend to stem from incorrect use of the hot plates, lack of concentration when setting up a reaction using a hot plate, or lack of cleaning of the hot plate. The hot plate should regularly be tested for normal function and that can, e.g., be done by heating water on it and keeping an eye out on the temperature and settings. Under certain circumstances, it can be good to have the hot plate secured on a lab jack (that needs to be large enough to provide secure support), so that in the event of overheating, it can easily be lowered to remove the heat source.

Thermometers that use mercury should be made extinct from laboratories today and instead you should use, e.g., an alcohol-based thermometer or a digital one if you need to measure the temperature.

Always use a ***stirring magnet*** or ***boiling stones*** when heating liquids, as these help prevent superheating and yield gentler boiling of the liquid with less bumping. When you set up a long reaction using a hot plate, make sure to keep an eye out on it until the reaction has reached its maximum temperature and is steady, to ensure that it can progress safely. Make sure that the surrounding environment is also safe and that there are no solvents or chemicals close by that can be hazardous should anything malfunction.

Water baths are used for temperatures in the range of 0−80°C. *Oil baths* are recommended for temperatures up to 200°C and if there is a need for even higher temperatures, a dry sand bath can be used. Other options are also possible, such as aluminum plates specifically designed for the size of the reaction vessel to be used. When heating a mixture, it is important to use the right glassware (e.g., Pyrex) and make sure that the glassware is not cracked or anything like that. Thick bottomed glassware (e.g., suction flasks) should also not be used on hot plates.

Ultra sonic baths or *sonicators* are used for cleaning, disintegration of compounds and cells, or to help solvate chemicals, to name a few of their uses. These sonic baths produce a high frequency sound in the range of 16,000−40,000 Hz. It can be advisable to use ear protection when using these sonic baths, since this frequency of sound can cause hearing damage (20,000 Hz from a sonic bath is equivalent to 100 dB). It is also recommended that sonic baths be set up in a way that minimize chances of anyone playing around with it. If a finger or a hand is inserted into a sonic bath, one can expect bad tissue damage.

Lyophilizers (freeze dryers) produce vacuum and thus lower the boiling point of water so that it can boil at room temperature. It is important to learn to work with and operate the lyophilizer before changing any settings or moving cables around or the pump that yields the vacuum. Carelessness or rashness (e.g., during connection or disconnection) can lead to a flask that is connected to the instrument breaking because of the great vacuum that the instrument produces. The pressure-depression is so much that shards of glass could fly in all directions. Therefore, it is important that the flask that is to be connected to the instrument be strong enough to withstand the vacuum. Flasks and other glassware that have any cracks in them are, in general, not likely to endure this type of vacuum. You should always use safety glasses when working in the vicinity of lyophilizers. The same applies to *rotavapors*. Be careful to use whole flasks (no cracks) when using rotavapors because of the reduced pressure being produced in those systems.

Centrifuges are categorized depending on how fast they rotate and whether they have an internal cooling system or not. There are generally three types:

- *Low speed centrifuges* do not go over 5000 rpm. Many benchtop centrifuges fall under that category, but those are generally small centrifuges that are easy to move between places.
- *High speed centrifuges* can go up to 25,000 rpm. These centrifuges are usually freestanding on the floor because of how great the rotational speed is and the instrument needs to be able to stand on stable ground.
- The third category is *ultracentrifuge* that can go over 100,000 rpm. In light of the great rotational speed of those centrifuges, you need to be really careful around them and it is important to adopt very precise and meticulous working procedures when samples are located in the instrument.

These centrifuges can cause extreme hazards if they are used improperly. Centrifuges should be checked regularly. The most common mistake when using a centrifuge is to forget to balance the samples that are put in the centrifuge. Vials and glassware being used should be whole and without any cracks so that they do not break from the great centrifugal force they experience in the instrument. You should never overfill any vials that go into the centrifuge. Do not fill more than approximately 3/4 of the vial and close it securely before starting the instrument. Open vials may result in droplets formed inside the instrument that contaminates everything in there and can, with time, cause erosion to parts of the centrifuge. All vials should be weighed and equilibrated before starting the centrifuge. Although a weight difference of 0.5 g might *seem* innocent, this difference is analogous to a 25-kg weight difference on each side of the axis when the rotational speed is up to 50,000 rpm!

Autoclaves are primarily found in laboratories where people work with microorganisms (bacteria, virus, fungi, parasites, prions, etc.) and other biological samples and hazards, and also in laboratories where sterile tools, products, formulations, drugs, or chemicals are needed, and to set up certain chemical reactions. These instruments are sort of pressure chambers that heat water up to 120°C. That results in a significant pressure that is used to sterilize tools or solutions but can also be used to initiate certain reactions. Because of the increased pressure, no one should use an autoclave except after receiving proper training regarding the use of the instrument. It is recommended that only borosilicate glassware be used in autoclaves, i.e., Pyrex or Kimax, as they can withstand the pressure and temperatures used. The instrument should not be started unless it is securely closed, and all safety locks have been checked. You should never put erosive chemicals (acids or bases), flammable chemicals (e.g., methanol, ethanol, chloroform), plastic that can melt, phenols, or radioactive chemicals in an autoclave.

When the pressure chamber is opened, it should be done carefully (it is good to wait about 10 min before it is opened). Keep your face and hands away from the opening so that you do not get hot steam at you and use gloves.

Noise in laboratories

All noise, buzzing sounds or such in the work environment can have negative impact on your work, especially that which requires focus, whether it is computer work or laboratory work. The psychological effect of noise pollution can be significant and result in stress, poor communications, frustration, and even anger. Additionally, noise pollution can increase risk of accidents. A good frame of reference is that noise should not exceed 80 dB. If the noise exceeds that level, ear protection is recommended. The iPod generation is currently working in the laboratory as well and is used to having music constantly in its ears. It is important though for everyone to be able to hear easily and right away what is happening around them, especially if warnings need to be made quickly, so you should not be listening to anything in both ears that might close you off from your surroundings. You should also make sure that if you are listening to anything, that it does not reduce your concentration.

Waves (microwaves, radiowaves, and light)

If possible, avoid spending prolonged periods of time around *microwaves* or *radiowaves*. The eye is sensitive to all radiation. It is important to keep these waves at low intensities or within the appropriate instruments and blocked of from escaping the room.

Ultraviolet light can cause skin burns and eye damage (e.g., photokeratitis) but the effect is dependent on the wavelength of the light that you are working with. It is recommended to limit use of wavelengths <320 nm, or at least shield the light properly.

Infrared light can also cause damage to skin and eyes, as it can increase the temperature of the skin/internal temperature of the eye. You should always use proper PPE when working with these types of lights.

Laser is high energy light that is categorized by its power (measured in W or W/m^2). The classes are five: 1, 2, 3a, 3b, and 4. When working with lasers, it is important to know their direction and the power of the laser. Anything that can reflect light, e.g., mirrors, should not be unnecessarily in the working area. Prepare yourself well and protect your eyes especially when working with strong laser beams (Fig. 3.5a).

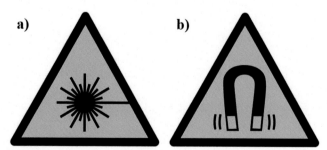

Figure 3.5 a) Hazard sign for a lab where lasers are being used. b) Hazard sign indicating a strong magnetic field. *(Published with the permission of the International Organization for Standardization, ISO.)*

Strong magnets, such as ones found in NMR and MRI instruments can also be hazardous. It is important to keep loose metal objects away from the strong magnetic field as they can suddenly be pulled quickly to the magnet, creating a potentially hazardous situation. People with cardiac pacemakers, implantable cardioverter defibrillators (ICDs), and other implants that might be affected by the magnetic field should also stay away from strong magnets, as the magnetic field can affect, or even inhibit, how these implants operate, which could result in death or serious injury to the user. Lastly, be careful about bringing electronic objects or other magnetic media, such as credit cards, too close to strong magnets, as these may be affected and damaged if they get too close to the magnet. Where there are strong magnets, it is good practice, and often required, to have the 5 gauss line clearly labeled (equivalent to 5×10^{-4} Tesla), and the appropriate hazard signs clearly visible (Fig. 3.5b).

Proper protocols are the key to success

A good preparation is the foundation for things to work out and the use of proper protocols are the key to success. If there is any doubt about the execution of a project, contact your supervisor or someone who knows what to do and get a demonstration or guidance. Here below are a few things that are worth keeping in mind when starting a research project:

- When in doubt, ask your supervisor and/or a coworker for information and a demonstration/training on how to use the instrument or chemical that you are working with.
- Never use an instrument for a different purpose than the one it was intended for.
- Never walk away from an open flame or a burner.
- Treat all chemicals with the utmost carefulness.
- Do not touch anything in the laboratory that you do not know what is or what it is used for.
- If you put your compound in a container, label it immediately and store it properly.
- Always use personal protective equipment correctly and put long hair in a ponytail.
- Always use safety glasses/goggles and the right type of gloves.
- Always be neat and put everything properly away when not in use.
- Wash your hands regularly and always when you leave the laboratory.
- Never work alone in the laboratory.
- In the event of an accident, call for help immediately and report the accident.

CHAPTER 4

Chemicals

Chemical accidents in research labs are often due to people underestimating their circumstances or not doing adequate research on the chemical properties before starting their work. This *underestimation* can then lead to unexpected events or even reactions causing an accident. In chemistry people work with a variety of chemicals in all possible states of matter, solids, liquids, and gas. Gases are usually compressed at high pressures in appropriate gas tanks. The chemicals are then organized into different hazard groups depending on their properties and behavior. These categories include flammable, explosive, oxidizing, corrosive, toxic, reactive, and radioactive compounds. It is important to be aware of what kind of chemicals you are working with as well as what kind of chemicals are being used in your work environment. In many chemical labs you can find examples of chemicals from most, if not all, of these hazard *categories*.

Taste, smell, and texture

Before we go on, it is worth repeating from earlier that it is strictly forbidden to smell, taste, or touch the chemicals you are working with. Several decades ago, pharmaceutical identity tests were based on the above-mentioned criteria, i.e., smell, taste, and feel. They were called organoleptic tests. Today, we do not use organoleptic tests at all. It is simply not worth it. We have a number of analytical instruments and methods that are better suited for identification and there are also just way too many hazardous compounds that we tend to work with. It also used to be that people pipetted liquids using the mouth, but now we have tools for that so you should never suck a fluid up in lab using your mouth.

If you *have to* smell a sample, the recommended method is to use your hand to waft the air toward your face in order to catch a whiff of the chemical. But in general, it is much better to avoid smelling the compounds, especially if you are not familiar with them. When working with especially volatile compounds, it is important to keep them in a well-ventilated fume hood (Fig. 4.1).

Handbook for Laboratory Safety
ISBN 978-0-323-99320-3
https://doi.org/10.1016/B978-0-323-99320-3.00008-2

Figure 4.1 You should never taste, smell, or touch the chemicals you are working with. *(Credit: Sveinbjörn Gizurarson).*

Importance of preparation

Research and experiments can be very enjoyable, but they require us to be alert and organized in our work. Carelessness or even just a single moment of thoughtlessness could become drastic and cause a lot of damage, both to ourselves and to our environment. If we are well prepared, the chances of unexpected accidents are reduced. It is therefore important to carefully read all protocols and prepare yourself well before starting an experiment or a new project.

The first step in this preparation is to familiarize oneself with the proper methods and protocols for handling the chemicals that you are planning to work with. Take some time to learn about the chemicals, including reading about the relevant safety considerations, PPE requirements, and what to do in the event of a spill or an accident. It is also useful to read about the physical properties of the materials, such as their boiling point, flash point, and vapor pressure. This information should be available in the safety data sheet (SDS) for the chemical and is even sometimes on the product label. Other useful information to keep in mind are: what hazard class(es) does this compound fall into? How reactive is it? Can it make peroxides? Is it an oxidizer or reducer? Does it react with water? Does it dissociate easily? Is it sensitive to light, heat, or oxygen from the atmosphere? All of these are useful questions to ask yourself before you set up a new or a scaled-up reaction.

You should not forget either to look at the health and environmental effects of the chemicals you are going to use: Are they corrosive, toxic, or hazardous to the environment? Can they cause/induce allergies or allergic reactions? Are they carcinogenic, or can they cause mutations or birth defects? Can they have acute or chronic toxic effect? Are there any specific requirements concerning the handling of or use of these chemicals? Is their use restricted in any way? Do you need special training to handle the chemicals? Do you need to have an antidote close by when you are working with the compound (e.g., for cyanide or hydrofluoric acid)? For common compounds, most of this information should be available. Later in this chapter, we will discuss in more detail how to access it in a quick and simple way.

When you are preparing to run chemical reactions, there are a few more things to consider. What side products might be formed? How safe are these side products? Are any of them gases? Is the reaction likely to be exothermic or endothermic? How much energy might be released, and should you take steps to cool it down so that it does not overheat? Are there any chemicals that should not be in the proximity when the reaction is run? Can the reaction be dangerous for other people working in the laboratory? What might go wrong? How closely do I have to monitor the reaction? Do I need to be present while it runs, or can I go grab some coffee, or leave it running overnight? It is important to keep asking yourself these types of questions and always think about the safety of your lab work. When in doubt, discuss it with your coworkers or supervisor! A good preparation is important and can often even save time. The key to good results is a well and thoroughly thought-out experiment.

The following sections discuss where to find the relevant safety information you need to prepare yourself properly for experiments involving chemicals. In that context, the SDSs are discussed along with chemical labels. Following that, we touch on the main hazard classes, and provide some general guidelines about what to be aware of when working with chemicals in those groups. These guidelines are by no means exhaustive, but will hopefully prove useful as primary considerations when working with these chemicals. Remember, it is ultimately your responsibility to look up the relevant safety information relating to the specific chemicals you are working with.

Safety data sheets

For many of the questions that have been posed here above, the answers are commonly found in the SDS that the chemical provider should provide along with the chemical or at least have accessible on their website. In most countries, the vendors are legally required to provide these either physically or online, so if you encounter any troubles in finding these, you should be able to contact the vendor. You can often find the SDS from other sources online as well.

The SDS includes information on the chemical, its behavior, proper handling of it, what personal protective equipment to use, what to do in the event of a chemical spill, personal exposure, etc. Remember to read the SDS over *before* you start working with a new compound and know where you can find it if you need to refer to it again later. It is often good to have a folder and/or printouts of the SDSs for the chemicals that you have in the laboratory so that it is easy to review information quickly about specific chemicals that you are planning to use.

The structure of SDSs has been standardized internationally by the Globally Harmonized System of Classification and Labelling of Chemicals (GHS), making it easy to look up specific information quickly, regardless of which vendor you purchased the chemical from. A list of the sections that should be included in SDSs can be found in Table 4.1.

It is important to remember that it is your responsibility to familiarize yourself with the SDSs for the compounds you are working with. The more SDSs you read through, the faster you will be at finding the main information you need, and you will find that some sections may be more relevant for you than others. It is useful to start with Section 2, looking especially at the hazard and precautionary statements there to get a quick overview of the hazards attached to the compound. When you first receive a new chemical, it is also useful to look at Section 7.2 to find the appropriate storage location for the product. Under Section 7 you will also find information about the safe handling of the compound (7.1) and in Section 8.2 there is important information about what PPE you should wear while working with the compound. Make sure to wear the appropriate PPE!

Before using the chemical in a reaction, you should also read through Section 10 to make sure that your reaction conditions do not create avoidable hazards, e.g., from possible hazardous reactions. If there are unavoidable hazards accompanying your planned experiment, make sure to take the appropriate precautionary steps and consult a coworker and/or your supervisor before setting the experiment up. Maybe it turns out that the hazardous reaction is actually avoidable.

In case of an accident and chemical exposure, it is important that you have looked through the first aid measures, listed in Section 4, *before-hand.*

In the event of an accident, you do not want to have to start by searching for the necessary first aid measures, but rather be able to act on them right away. Section 5, about firefighting measures, also includes important information that you need to know *before-hand*, in the event of an accident involving fire. This information should also be available to firefighters who may need to respond to an accident in your lab, as they need to know what hazards they need to be ready for and what approach might work best to extinguish the fire.

Table 4.1 Overview of the structure of safety data sheets (SDS).

1. Identification of the substance/mixture and of the supplier
   1.1. Product identifier
   1.2. Relevant identified uses of the substance or mixture and uses advised against
   1.3. Details of the supplier of the safety data sheet
   1.4. Emergency telephone number
2. Hazard identification
   2.1. Classification of the substance or mixture
   2.2. Label elements
   2.3. Other hazards
3. Composition/information on ingredients
   3.1. Substances
   3.2. Mixtures
4. First aid measures
   4.1. Description of first aid measures
   4.2. Most important symptoms and effects, both acute and delayed
   4.3. Indication of any immediate medical attention and special treatment needed
5. Firefighting measures
   5.1. Extinguishing media
   5.2. Special hazards arising from the substance or mixture
   5.3. Special protective equipment and precautions for firefighters
6. Accidental release measure
   6.1. Personal precautions, protective equipment, and emergency procedures
   6.2. Environmental precautions
   6.3. Methods and material for containment and cleaning up

(Continued)

Table 4.1 Overview of the structure of safety data sheets (SDS).—cont'd

7. Handling and storage
   7.1. Precautions for safe handling
   7.2. Conditions for safe storage, including incompatibilities
8. Exposure controls/personal protection
   8.1. Control parameters
   8.2. Appropriate engineering controls
   8.3. Individual protection measures, such as PPE
9. Physical and chemical properties
   9.1. Information on basic physical and chemical properties
   9.2. Other information
10. Stability and reactivity
    10.1. Reactivity
    10.2. Chemical stability
    10.3. Possibility of hazardous reactions
    10.4. Conditions to avoid
    10.5. Incompatible materials
    10.6. Hazardous decomposition products
11. Toxicological information
    11.1. Information on toxicological effects
    11.2. Additional information
12. Ecological information
    12.1. Ecotoxicity
    12.2. Persistence and degradability
    12.3. Bioaccumulative potential
    12.4. Mobility in soil
    12.5. Other adverse effects
13. Disposal considerations
    13.1. Waste treatment methods
14. Transport information
    14.1. UN number
    14.2. UN proper shipping name
    14.3. Transport hazard class(es)
    14.4. Packing group
    14.5. Environmental hazards
    14.6. Transport in bulk according to IMO instruments
    14.7. Special precautions for user
15. Regulatory information
    15.1. Safety, health, and environmental regulations/legislation specific for the substance or mixture
    15.2. Chemical safety assessment
16. Other information

Adapted, with permission from the UN, from the GHS, 9th ed.).

Section 11 contains toxicological information, including the lethal dose 50 value (LD_{50}) where it is known. The LD_{50} value is the amount needed to kill half of the test population. These values are given specific to the animals they were tested on and tend to be given as mass of compound per kg of animal. It can therefore be used as a rough estimate for how toxic the compound is. By calculating the LD_{50} value for you, based on your weight, you can evaluate how much of the substance you are willing to work with at a time. The further you are from the LD_{50} value for your weight, the better. It is strongly discouraged to work with amounts that would constitute the LD_{50} value for you or higher, if you can avoid it.

There is more useful information included in the SDSs, so make sure to read through them for the chemicals you will be working with. As you read more and more SDSs, it will eventually be easier and easier to quickly look up the relevant information you need to cover to work safely with the compounds.

Chemical labels

Chemical labels should be clear and follow international rules and regulations. Manufacturers and importers are responsible for all labels being correct. The labels should include a clear product identifier, often including the batch number, and information about the manufacturer, and/or importer. For hazardous chemicals, the labels should also include a signal word along with pictogram(s), hazard statement(s), and precautionary statement(s) (Fig. 4.2). Basically, the label should allow for a quick guide to the main hazards associated with the chemical along with information about the main precautions that should be taken while working with it. The label is not a complete substitute for the SDS though.

As the age of the chemicals can affect their quality and, in some cases, their safety, it is a good practice to label new chemicals with the date you receive them as well as the date when they are first open. This is mandatory in many companies, especially in the pharmaceutical industry but is also generally a good practice from a safety perspective.

It should be noted that the pictograms (Fig. 4.3) and rules and regulations have changed somewhat in the last couple of decades. If you have very old chemicals in your laboratory, be mindful of that and look up the old pictograms for reference. It also used to be enough to include the codes for the hazard and precautionary statements, rather than the actual statements.

Figure 4.2 An example of the label of a chemical container. 1: Product name; 2: Product number; 3: Pack size; 4: Batch number; 5: Expiration date; 6: Composition; 7: Use; 8: CAS number; 9: EC number; 10: Density; 11: Molar mass; 12: Product specifications; 13: UN Code; 14: Signal word; 15: Hazard and Precautionary statements; 16: Hazard pictograms; 17: Country of origin; 18: Vendor's contact details. *(With permission from Chem-Lab NV).*

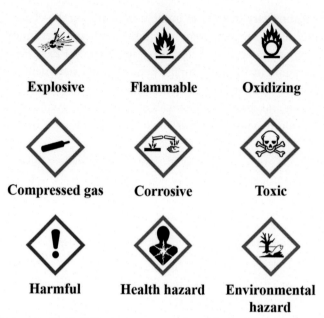

Figure 4.3 Standardized GHS pictograms used on labels. *(With permission from the United Nations, from the GHS, 9th ed).*

Thankfully, this has been updated so people do not have to memorize what the codes represent or look them up especially. These codes are still used though, e.g., in the SDSs, and they provide important references for translations, for example.

The hazard pictograms indicate the kind of danger that the chemical can cause. They show the type of hazard, e.g., flammable or corrosive. There are nine hazard pictograms, and the same compound can be assigned more than one of those.

The hazard statements give us more specific information about the type of hazard this chemical can cause, e.g., "eye irritant." The hazard statements are categorized into physical hazards (H2xx codes), health hazards (H3xx codes), and environmental hazards (H4xx codes). A few examples of these hazard statements are

H224   Extremely flammable liquid and vapor.
H240   Heating may cause an explosion.
H317   May cause an allergic skin reaction.
H330   Fatal if inhaled.

The precautionary statements guide us in handling the chemicals. There are five precautionary statement categories: general precautionary statements (P1xx), prevention precautionary statements (P2xx), response precautionary statements (P3xx), storage precautionary statements (P4xx), and disposal precautionary statements (P5xx). This information should guide us in how we use and store the chemicals in addition to giving brief and clear information about what to watch out for, as can be seen in the examples here below.

P202   Do not handle until all safety precautions have been read and understood.
P232   Protect from moisture.
P235   Keep cool.
P243   Take precautionary measures against static discharge.

The full list of hazard and precautionary statements can be found in Table 4.8 at the end of this chapter.

Toxic compounds

Many compounds are harmful to our health and have turned out to be dangerous for both men, animals, and nature. Many studies are behind these types of categorizations. Some chemicals are so harmful that governments have established specific regulations regarding their use. Among these are various pharmaceutically active compounds and radioactive chemicals to name a couple of general examples. Although a chemical can be harmful, it can still be used for medicine and research, but their handling must be done by individuals who have received the necessary training to handle them in a safe manner. It is the responsibility of supervisors to ensure that whoever works with hazardous chemicals receives the proper training and skill to handle the chemicals *before* they start using it in their work.

Hazardous chemicals are categorized by how much or how little of the compound it takes to cause real harm. This categorization uses the acute toxicity values, with the current criteria for the different categories being as indicated in Table 4.2. The values in the table refer to the acute toxicity estimates that are often based on the LD_{50} and LC_{50} (lethal concentration for 50% of test cases) values where available. For compounds in categories 1–3, the skull and crossbones hazard pictogram should be used. For category 4 compounds, the exclamation mark symbol is used, and no symbol is used for category 5 compounds.

Table 4.2 Acute toxicity estimates based on exposure route.

Exposure route	Category 1	Category 2	Category 3	Category 4	Category 5
Oral (mg/kg)	≤ 5	5–50	50–300	300–2000	2000–5000
Dermal (mg/kg)	≤ 50	50–200	200–1000	1000–2000	See criteria
Gases (ppmV)	≤ 100	100–500	500–2500	2500–20,000	in footnote[a]
Vapors (mg/L)	≤ 0.5	0.5–2.0	2.0–10.0	10.0–20.0	
Dusts and mists (mg/L)	≤ 0.05	0.05–0.5	0.5–1.0	1.0–5.0	

[a]Compounds are classified into category 5 if reliable evidence exists that the LD_{50}/LC_{50} values are above the category 4 values, and thus, assignment into one of the earlier categories is not warranted. This is sometimes done through extrapolation of data or estimation.

For novel compounds, the necessary toxicity values to categorize them do not exist. Therefore, novel compounds should always be treated with utmost care, as they might even turn out to have category 1 toxicity. This is especially important in chemical research where you might be the first one in the world to make some of the compounds that you will be working with.

For some toxic compounds, antidotes exist. If that is the case, you should make sure it is close by at a known location, before starting to work with that hazardous compound. Hydrogen fluoride (HF) is an example of a compound that you should not work with unless the antidote is close by. It is a dangerous contact poison, with effects that do not always appear immediately. Thankfully, HF antidote gels exist that should be used immediately in the event of exposure to reduce the damage. We cannot assume however that antidotes are available for all toxic compounds or that it provides guaranteed protection against possible toxic effects. In some cases, the antidote may even be dangerous itself, and require proper training to handle.

We also need to realize that the toxicological effects of chemicals are not always acute or visible right away. They might even surface several months or even many years later. Many dangerous chemicals are odorless, colorless, dissolve easily, and can *diffuse* rapidly through skin. For compounds with known adverse effects, e.g., carcinogens and chemicals with reproductive toxicity or target organ toxicity, the appropriate health hazard symbols should be used.

Organic solvents are examples of dangerous chemicals. Most of them are flammable and can also cause drowsiness. Chlorinated solvents can affect the cardiovascular system but solvents that include other halogens can form very toxic gases, e.g., when burned. This should be kept in mind when preparing experiments that use these solvents.

In general, it is always best to limit your exposure to all chemicals in the laboratory, regardless of their known toxicity. Good work practices, vigilance, and organization are often what characterize a successful experiment.

Toxicity assessment for chemical mixtures

It should be relatively straight-forward to access information about the hazards associated with specific chemicals through their SDS. The same cannot be said about chemical mixtures. When working with mixtures, researchers might need to assess the danger themselves based on what chemicals are in the mixtures. This type of hazard assessment is built on calculations where the percentage of each compound is evaluated in the overall mixture. Let us use the following example to demonstrate how these calculations are done:

Example:

A research lab worker prepares a mobile phase for an HPLC instrument. The solution is composed of the following chemicals:

33% Acetonitrile (Group 3: LD_{50} = 269 mg/kg)

1% Tetrahydrofuran (Group 4: LD_{50} = 1,650 mg/kg)

66% Water (Group 5: LD_{50} = >90,000 mg/kg)

This hazard assessment is done by calculating all chemicals with LD_{50} under 2000 mg/kg (Categories 1–4). That means that we do not need to consider the water in our calculations. We use the following equation for that:

$$ATE_{mix} = \frac{100}{\sum_i \frac{C_i}{ATE_i}}$$

where ATE_{mix} is the *acute toxicity estimate* of the mixture, C_i is the percentage of each chemical in the mixture, and ATE_i is the respective acute toxicity estimate for each chemical in the mixture. For the solution described above, the ATE_{mix} would be

$$ATE_{mix} = \frac{100}{\frac{33}{269} + \frac{1}{1650}} = 811 \text{ mg/kg}$$

Example:—cont'd

The solution described above has the ATE_{mix} value of 811 mg/kg which means that the final mixture falls under category 4, being hazardous to our health and it should be labeled accordingly.

Corrosives—acids and bases

Most, although not all, corrosives are either acids are bases. As usual, make sure to wear the appropriate PPE when working with these, and work inside a well-ventilated fume hood, especially when working with corrosive gases. Be careful when preparing acidic or basic solutions and when mixing them together. The reactions between them tend to be exothermic and, in some instances, they can also result in gas formation. Because of the exothermicity of the reactions between acids and bases, they need to be stored separately.

When preparing acidic solutions, remember to add the acid to the water. Otherwise, the exothermicity of the mixing process can result in the acid splashing up and if you become the victim of a splash, it can cause acid burns. In general, we want to start with a neutral solution and slowly move away from that by adding the acid to it rather than the other way around.

During reaction work-up with either acid or base, using a separatory funnel, make sure to allow the mixture to vent frequently to prevent pressure build-up. These washing procedures frequently include gas formation, making it vital to release the pressure in a safe manner, rather than let the pressure build-up force its own way out, with the accompanying chemical spill and the undesired potential chemical exposure.

In the event of a spill on yourself, take off any contaminated clothing immediately and wash the exposed area extensively with water. Depending on the size of the spill and location, you may need to use the emergency shower. You may need to disrobe completely so that the corrosives do not get a chance to be "trapped" in contaminated clothing that remains on your body. There are examples where modesty following a serious chemical accident led to chemical burns in private areas.

Acid exposure generally leads to a clearly felt chemical burn, but exposure to bases can be more subtle, although it is also very serious and can reach deep into the tissue. Be cautious and alert so that you can take the appropriate response to a chemical exposure right away, regardless of whether you feel any immediate pain from the exposure.

Naked and shocked

In a European pharmaceutical manufacturing plant, a technician was asked to make a diluted sulfuric acid solution. He measured the required amount of the acid into a 1 L Erlenmeyer bottle and then he added the water. A few seconds later the water started overheating vigorously resulting in an explosion!

The technician ran for his life, under the emergency shower and removed all his clothes. He was really shocked. Something that looked so simple turned out to be life-threatening. His colleagues came over to assist him and get him dry clothes. When cleaning the area the acid had stained the ceiling, the table and there were burns on the floor. His clothes were almost dissolved by the acid. There were huge holes everywhere. But the technician was lucky.

Gases

Most laboratories need to use some gases in their research. The same is true for many analytical instruments and other equipment. Therefore, you can often find various sizes of gas cylinders within the lab. Smaller gas cylinders tend to be used for specific experiments, whereas bigger gas cylinders are used for larger instruments as well as being connected to Schlenck lines in fume hoods. Nitrogen (N_2) and Argon (Ar) are among the more common gases used, allowing us to run experiments under an inert atmosphere. In most of these cases, the gas is stored under a lot of pressure and so we need to be careful with how we store and work with (compressed) gases and gas cylinders. Chapter 3 contained a discussion about the appropriate storage of gas cylinders, so start by making sure that your gas storage conforms with the safe storage requirements. Lines are often set up from the gas storage location to transport the gas to where it will be used. It is important to note that there may be different tube materials for each type of gas, since there is different compatibility between the gas and the tube material. In Table 4.3 there is a list of compatibility or resistance of gas materials and the tube materials.

Table 4.3 The compatibility and resistance of gas materials and tube materials. *F*, Fair; *G*, Good; *NR*, Not recommended.

Compound	Al	Bs	Bu	CSt	Cu	NiCu	Np	Nr	Ny	PE	PVC	SS	Te
Acetylene	G	F	G	G	NR	NR	F	NR	F	NR	G	G	F
Air	G	G	G	G	G	G	G	G	G	G	G	G	G
Aminomethane	F	NR	F	G	NR	G	NR	F	G	G	G	G	G
Ammonia	F	NR	G	G	F	NR	F	F	NR	–	G	F	G
Argon	G	G	G	G	G	G	G	G	G	G	G	G	G
Arsine	F	F	G	F	G	G	F	G	G	G	G	F	F
Boron trichloride	NR	F	NR	G	NR	G	NR	NR	NR	F	F	G	G
Boron trifluoride	NR	NR	F	F	NR	F	F	NR	NR	F	F	F	F
Bromoethane (methyl–)	NR	G	NR	F	G	G	NR	NR	–	F	F	G	G
Bromoethene (vinyl–)	NR	NR	F	F	NR	G	NR	–	F	F	F	G	G
1,2-Butadiene	G	G	NR	G	G	G	F	NR	F	F	F	G	G
n-Butane	G	G	NR	G	G	G	F	G	G	NR	G	G	G
iso-Butane	G	G	NR	G	G	G	F	G	F	NR	G	G	G
1-Butene	G	G	NR	G	G	G	NR	G	F	NR	G	G	G
cis-2-Butene	G	G	NR	G	G	G	F	G	G	NR	G	G	G
iso-Butene	G	G	NR	G	G	G	F	G	G	F	G	G	G
trans-2-Butene	G	G	NR	G	G	G	F	G	G	NR	G	G	G
1-Butyne	G	NR	–	G	NR	–	–	–	–	–	–	G	NR
Carbon dioxide	G	G	F	G	G	G	NR	NR	G	F	G	G	G
Carbon monoxide	G	G	NR	F	G	G	F	F	–	F	F	G	G
Carbon oxyfluoride	–	G	–	G	G	G	–	–	–	–	–	G	–
Carbonyl sulfide	F	F	NR	F	F	F	F	NR	G	NR	G	NR	G
Chlorine	NR	F	NR	F	F	NR	NR	NR	NR	NR	F	G	G
Chlorodifluoroethane	G	G	F	G	G	G	G	F	G	F	F	G	G

(Continued)

Table 4.3 The compatibility and resistance of gas materials and tube materials. *F*, Fair; *G*, Good; *NR*, Not recommended.—cont'd

Compound	Al	Bs	Bu	CSt	Cu	NiCu	Np	Nr	Ny	PE	PVC	SS	Te
Chlorodifluoromethane	G	G	F	G	G	G	G	NR	NR	F	F	G	NR
Chloroethane	F	F	G	F	F	F	F	G	—	—	NR	G	G
Chloroethene	NR	F	F	F	NR	F	G	—	G	G	—	F	G
Chloromethane	NR	G	NR	F	G	F	NR	NR	—	NR	NR	F	G
Chloropentafluoroethane	G	G	G	G	G	G	G	G	F	G	G	G	F
Cyanic chloride	—	—	—	NR	G	—	NR	NR	—	—	F	G	—
Cyclopropane	G	G	NR	G	G	G	F	G	—	NR	G	G	G
Deuterium	G	G	G	G	G	G	G	G	G	G	G	G	G
Diborane	G	G	G	G	G	G	G	G	F	F	—	G	G
Dichlorodifluoromethane	G	G	NR	G	G	G	G	F	G	F	G	G	NR
Dichlorofluoromethane	G	G	NR	G	G	NR	NR	NR	G	NR	G	G	G
Dichlorosilane	NR	F	NR	F	F	G	NR	NR	NR	F	—	G	G
1,2-Dichlorotetrafluoroethane	G	G	F	G	G	G	G	G	G	G	G	G	G
1,1-Difluoroethane	G	G	F	G	G	G	F	G	G	F	G	G	G
1,1-Difluoroethene	G	F	—	G	NR	G	—	—	—	—	—	G	G
Difluoromethane	G	G	G	G	—	F	G	—	—	G	—	G	G
Dimethylamine	F	NR	F	G	NR	NR	NR	NR	—	G	G	G	G
2,2-Dimethylpropane	G	G	F	G	G	G	F	G	—	NR	G	G	G
Epoxyethane	G	G	NR	F	NR	G	NR	NR	—	F	F	G	G
Ethanamine	F	NR	F	G	NR	F	F	NR	NR	G	G	G	G
Ethane	G	G	NR	G	G	G	G	G	G	F	G	G	G
Ethanedinitrile	G	NR	NR	F	NR	—	G	NR	—	G	G	G	G
Ethene	G	G	F	G	G	G	G	G	G	F	G	G	G
Fluorine	F	F	NR	F	F	G	NR	NR	NR	NR	NR	F	F
Fluoromethane	G	G	G	G	G	G	G	G	G	G	G	G	G

Chemical	1	2	3	4	5	6	7	8	9	10	11	12	13
Helium	G	G	G	G	G	G	G	G	G	G	G	G	G
Hexafluoroethane	G	G	—	G	G	G	G	G	G	G	G	G	NR
Hydrogen	G	G	G	G	G	G	G	G	G	G	G	G	G
Hydrogen bromide	G	F	G	G	NR	NR	NR	NR	NR	NR	NR	NR	NR
Hydrogen chloride	F	F	F	F	NR	NR	NR	F	NR	F	NR	NR	NR
Hydrogen fluoride	F	F	NR	F	NR	NR	NR	NR	NR	NR	NR	NR	NR
Hydrogen iodide	G	F	NR	F	NR	NR	NR	F	NR	F	NR	NR	NR
Hydrogen sulfide	G	F	F	F	G	NR	F	G	F	F	F	F	F
Krypton	G	G	G	G	G	G	G	G	G	G	G	G	G
Methane	G	G	G	G	G	F	G	F	G	G	F	G	G
Methanethiol	G	F	—	F	F	F	F	NR	NR	G	NR	NR	F
Methoxyethene	G	G	G	G	G	G	G	G	G	G	G	G	G
Methoxymethane	F	F	G	F	—	F	F	F	G	G	G	F	F
Neon	G	G	G	G	G	G	G	G	G	G	G	G	G
Nitric oxide	G	G	G	G	G	F	G	G	G	G	NR	G	F
Nitrogen	G	G	G	G	G	G	G	G	G	G	G	G	G
Nitrogen dioxide	G	G	G	NR	NR	NR	NR	NR	NR	NR	NR	NR	F
Nitrogen trifluoride	G	G	NR	NR	—	NR	F	NR	F	NR	—	NR	G
Nitrous oxide	G	F	F	F	F	F	NR	G	G	G	G	F	F
Octafluoropropane	G	G	G	G	F	G	G	G	G	G	G	G	G
Oxygen	G	G	G	G	G	F	G	G	G	G	G	G	G
Phosgene	F	F	F	NR	NR	G	G	G	G	NR	NR	G	NR
Phosphine	F	F	—	F	F	F	F	G	F	F	G	G	F
Propadiene	G	—	—	G	G	G	G	G	G	G	G	G	G
Propane	G	NR	—	NR	G	G	NR	NR	G	G	G	G	G
Propene	G	F	G	F	NR	NR	F	G	G	G	NR	G	G

(Continued)

Table 4.3 The compatibility and resistance of gas materials and tube materials. F, Fair; G, Good; NR, Not recommended.—cont'd

Compound	Al	Bs	Bu	CSt	Cu	NiCu	Np	Nr	Ny	PE	PVC	SS	Te
Propyne	G	F	G	G	NR	NR	G	—	F	G	—	G	G
Silane	G	G	G	F	G	G	G	F	F	F	—	G	G
Silicon tetrachloride	NR	F	NR	F	F	NR	NR	NR	F	F	NR	F	G
Silicon tetrafluoride	F	F	NR	F	F	NR	NR	NR	NR	F	NR	F	G
Sulfur dioxide	F	F	—	F	F	F	NR	NR	NR	G	F	G	G
Sulfur hexafluoride	G	G	G	G	G	G	G	G	G	G	G	G	G
Tetrafluoroethane	—	G	—	G	G	—	—	—	—	—	—	G	—
Tetrafluoromethane	G	G	G	G	G	G	G	G	F	G	G	G	G
Trichlorosilane	F	F	NR	F	F	F	NR	NR	NR	—	—	G	G
Trifluoromethane	G	G	G	G	G	G	G	G	G	G	G	G	G
Trimethylamine	NR	NR	G	G	NR	G	F	F	NR	G	G	G	G
Xenon	G	G	G	G	G	G	G	G	G	G	G	G	G

Al, Aluminum; *Bs*, Brass; *Bu*, Butyl rubber; *CSt*, Carbon steel; *Cu*, Copper; *NiCu*, Nickel–Copper (Monel); *Np*, Neoprene; *Nr*, Nitrile rubber; *Ny*, Nylon; *PE*, Polyethylene; *PVC*, Polyvinyl chloride; *SS*, Stainless steel; *Te*, Teflon.

Different requirements apply to gas cylinders depending on the type of gas they contain. In the United States, for example, there are 45 types of fittings and regulators for gas cylinders depending on what type of gas they contain (CGA fittings). In Europe there are fewer of those. Make sure to look up what type of fittings and regulators are appropriate for your specific gas (Table 4.4).

Some valves are opened clockwise, while others are opened counterclockwise. If you have not received the proper training in working with gas cylinder fittings, contact your supervisor to request such training. Do not try to figure it all out yourself! All gas cylinders should be well labeled like other chemicals and pressure meters should be visible so that you can see how much of the gas is left in the container. It is important to check all valves, tubes, and detectors before opening the flow of gas. If something does not look right, close the valve immediately and ask for help. Empty containers should also be well labeled as such.

A common danger when working with gases in chemistry is when they are used along with liquid nitrogen, for example, when degassing a solution using a freeze–pump–thaw cycle. If a container is partly submerged under liquid nitrogen and open to a gas, such as nitrogen or argon, there is a chance of condensing some of the incoming gas. If this container is then closed and allowed to warm-up, the pressure builds up as the gas boils, which can easily result in an explosion. This can be even more dangerous if you have a liquid nitrogen trap for a Schlenck line that is left open to the atmosphere. There oxygen might condense in the trap which can result in an even worse explosion if it boils in a closed system. Liquid oxygen is blue, so if you ever see an unexpected blue color in your trap or container under liquid nitrogen cooling, take the appropriate precautions.

If gas is condensed, make sure to prioritize protecting yourself first, e.g., by setting up a blast shield right away. It is sometimes possible to prevent an explosion by cutting down on the flow of incoming gas and putting it under vacuum instead while keeping the condensed gas submerged in the liquid nitrogen bath. This allows the condensed gas to slowly evaporate again as the liquid nitrogen evaporates and the vacuum also helps boil it off faster. Still, make sure that taking these steps does not put you into greater danger *and make sure to notify anyone working close by of the danger.*

Table 4.4 Gas valves in the United States (CGA), Europe (DIN), and United Kingdom (BS) used for different types of gases.

Compound	United States (CGA)	Europe (DIN)	United Kingdom (BS)
Acetylene	510/300		4
Air	590/346	13	3
Allene (propadiene)	510	1	
Ammonia anhydrous	240	8	10
Argon	580	10	3
Arsine	350/660	5	4
Boron trichloride	660/329		
1,3-Butadiene	510	1	4
Butane	510	1	4
Butenes	510	1	4
Carbon dioxide	320	6	8
Carbon monoxide	350	5	4
Carbonyl fluoride	660/750	8	
Carbonyl sulfide	330	5	
Chlorine	660	8	6
Chloroethene (vinyl chloride)	290/510	5	7
Cyanogen (ethanedinitrile)	750/660	6	
Cyanogen chloride	750/661		
Cyclopropane	510		
Deuterium	350	1	1
Dimethylamine	705/240	5	11
Dimethyl ether	510	1	
Ethane	350	1	4
Ethylacetylene	510	1	
Ethylchloride	510/300	1	7
Ethylene	350	1	4
Ethylene oxide	510	1	7
Fluor	679/670		
Freon 14 (tetrafluoromethane)	320	6	6
Freon 22 (chlorodifluoromethane)	660/620	6	6
Helium	580/677	10	3
Hydrogen	350	1	4
Hydrogen bromide	330	8	
Hydrogen chloride	330	8	6
Hydrogen fluoride	330/660		
Hydrogen sulfide	330	5	15
Isobutane	510	1	4
Isobutylene	510	1	4
Krypton	590	10	3

Table 4.4 Gas valves in the United States (CGA), Europe (DIN), and United Kingdom (BS) used for different types of gases.—cont'd

Compound	United States (CGA)	Europe (DIN)	United Kingdom (BS)
Methane	350	1	4
Methyl chloride	660/510	5	7
Methyl mercaptan	330/750	5	
Monoethylamine	240/705	5	11
Monomethylamine	240/705	5	11
Mustard gas	750/350		
Natural gas	350/677	1	4
Neon	590/580	10	3
Nitric oxide	660/755	8	14
Nitrogen	580	10	3
Nitrogen dioxide	660/160	8	14
Nitrous oxide	326	6	13
Oxygen	540	6	3
Phosgene	660	8	6
Phosphine	660/350	5	4
Propane	510	1	4
Propylene	510	1	4
Silane	350/510	5	
Silicon tetrafluoride	330	8	
Sulfur dioxide	660/668	8	12
Sulfur hexafluoride	590/668	6	6
Trimethylamine	240/705	5	11
Vinyl chloride (chloroethene)	290/510	5	7
Xenon	580/677	10	3

If you are setting up a reaction that needs to be under gaseous pressure, the reaction vessel needs to be sturdy enough to withstand the pressure requirements. The appropriate vessel varies by the pressure needed. If the reaction is only at slightly elevated pressures and will be performed in a glass vessel, e.g., when running some hydrogenation experiments, you should double-check that there are no cracks visible in the glass. Pressurized reactions often require a sturdier reaction vessel though. This is often a metal pressure vessel that can be sealed tightly. Make sure to get the appropriate training for the specific pressure vessel you will be using, *before* starting to use it. It is also advisable to have some type of additional protective shield present when setting up a reaction under pressure and keeping it up while it is running in case something goes wrong mid-way through the experiment. At the end of the experiment, you will also need to show the utmost care when releasing any remaining pressure present. Again, make sure that you receive the proper training in handling the high-pressure vessels *before* you use them, and do the first experiment with them under direct supervision of an experienced researcher.

Some gases have additional hazards associated with them apart from just the pressure-related hazards. These additional hazards can include toxicity, like carbon monoxide, and flammability, such as hydrogen. In those cases, take extra precautions regarding those additional hazards. Unfortunately, serious gas-related explosions are still too common, with several of them resulting in deaths.

Hydrogen explosions[1]

In 2018, an experienced technician was setting up a cylinder with a mix of hydrogen and oxygen gases for an experiment that he was preparing. Hydrogen and oxygen are both flammable gasses and the reaction between them can be quite explosive. Simply mixing them together does not lead to a reaction though, but a spark or heat is needed. It is unclear what exactly went wrong this time, but sadly, during the experimental set-up, an explosion took place, resulting in the untimely death of the technician, and several other researchers being injured.

In a similar, tragic accident in 2019, a professor emeritus was performing a hydrogen-based test, that resulted in a serious explosion. The professor emeritus suffered serious wounds and burns from the explosion that eventually led to his death a couple of weeks after the accident.

[1] The first case was extracted from a news story from *The New Indian Express*, *IISc blast: Deceased was setting up cylinder, says survivor*. Published on December 12th, 2018. Accessed October 25th, 2021. https://www.newindianexpress.com/cities/bengaluru/2018/dec/12/deceased-was-setting-up-cylinder-says-survivor-1910361.html. The second case was extracted from a news story from *The Times of Israel*. Staff, T. *Technion professor dies after suffering serious injury in lab explosion*. Published October 26th, 2019. Accessed October 25th, 2021. https://www.timesofisrael.com/technion-professor-dies-after-suffering-serious-injury-in-lab-explosion/.

Explosives

Explosives can release a large amount of gases and/or heat very quickly when they explode, posing a serious threat to lives as this release contains a significant amount of destructive energy. If you need to work with explosives, request the appropriate training. Some explosives are well-known, like TNT and fireworks, and the proper precautions can then be taken from the start. These include storing the explosive safely away from potential ignition sources, including heat and direct sunlight. When working with these compounds, a good distance is also advisable, along with blast shields, and of course the appropriate PPE. Other specific precautions may depend on the specific explosive you will be working with, so make sure to get the appropriate training and do your own safety research, including reading the SDS carefully, before working with these compounds.

Another dangerous class of explosives is potentially explosive chemicals. These are chemicals that should not be explosive when you first receive them, but as time goes by, they can become explosive. An example of these is peroxidizable organic compounds, such as ethers. These pose a special danger because of the uncertainty about how dangerous they may be at a given time. For these compounds, it is also doubly important to label them clearly with the date received and date opened.

There is a lot of uncertainty regarding the safe handling of these potentially explosive chemicals, so again, it is important to do your own research for the specific compounds you are working with. A few general recommendations include storing these compounds as if they already are explosives, keeping them away from all ignition sources, including heat and direct sunlight. If you suspect that a container may be becoming over-pressurized, it might be good to release the pressure, but if you do that, wear all the necessary PPE, including protective heavy-duty gloves, and a safety glass screen between yourself and the container. Also consult a coworker and/or your supervisor before releasing the pressure.

Peroxide tests

In some cases, you can test the chemicals for explosive development. Peroxide tests include special dip strips to test for the semiquantitative detection of peroxides in organic and inorganic solutions. Other tests include the *ferrous thiocyanate method*, *iodide tests*, and a *titanium sulfate* test. It may also be possible to remove the peroxides that might have formed, so that is one consideration for a preventative measure.

Example of peroxide testing

There are different methods available to test your product for peroxides. The most easy one is to use peroxide test strips, a commercial test strip that can measure up to 25 ppm of peroxide in your product.

If you, however, want to quantify the peroxide in your product, you could use the following titration method that has been used in several pharmacopoeias:

(a) Weigh 5.0 g (±0.05) sample into a 250 ml glass Erlenmeyer flask.
(b) Add 30 ml of an acetic acidchloroform solution (60% acetic acid and 40% chloroform).
(c) Mix carefully until the sample is completely dissolved.
(d) Add 0.5 ml of saturated potassium iodide solution.
(e) Mix carefully for exactly one minute.
(f) Immediately add 30 ml of purified water, and shake well, to remove all iodine from the chloroform layer.
(g) Fill a burette with 0.1 N sodium thiosulfate.
(h) If the starting color of the solution is deep red orange, titrate slowly, with mixing until the color lightens. If the solution, however, has a light amber color, go directly to the next step.
(i) Add 1 ml of 1% starch solution (as an indicator).
(j) Titrate with sodium thiosulfate solution, until the blue gray color disappears in the aqueous (upper layer) and record the volume used.

You will also need to make a blank. The blank should contain all the reagents used, except the sample product, performed according to the protocol. Then to calculate the peroxide amount, use the following equation:

$$Peroxide value = \frac{(S - B) \times N_{Thiosulfate} \times 1000}{W_{sample}}$$

where S is the volume of thiosulfate used to titrate the sample and B is the volume of thiosulfate used to titrate the blank. $N_{Thiosulfate}$ is the exact strength of thiosulfate in N, and w_{sample} is the exact weight of the sample taken.

It is important to be aware of the age of the compounds and have a testing plan or a plan for the disposal of these potentially explosive compounds. It is especially important to test these chemicals before distillation as the distillation process creates high risk of explosion if peroxides are present, especially toward the end of distillation. If you do not know the age or origin of the container, do not test it but have it disposed of appropriately. If you observe crystals in the cap threads, take special caution. For old bottles, peroxides may have crystallized and can create a hazard when opening the bottle. In that case, it is better to have it disposed of directly without opening it at all. Table 4.5 includes a list of some known explosive and potentially explosive chemical families. If you are working with any of these, make sure to take the appropriate precautions and set-up a testing schedule where appropriate. In general, you should also be on the lookout for information in the SDSs about whether a compound is explosive or potentially explosive.

Explosive peroxides in an old ether bottle[2]

Several decades ago, researchers were sorting through some chemical bottles in a storeroom when they found two over 20-year old bottles labeled as containing isopropyl ether. Beneath the liquid layers in both bottles, there was a crystalline solid layer. Knowing that isopropyl ethers can form peroxides, and suspecting that the solid might be hazardous, they dispensed of the liquid and added water to the bottle. The solid proved to be insoluble in water. Eventually they took the bottles to a dump and tossed them as far away as possible. Then they threw stones at them to break them. The first stone that struck yielded a violent explosion, blasting mud and debris over the surroundings.

[2] This case was extracted from: Steere N.V. Control of hazards from peroxides in ethers. *J. Chem. Ed.* 1964, *41*, A575—A579.

Flammable compounds and oxidizers

The main precaution for flammables is to keep these chemicals from all ignition sources, such as open flames, hot surfaces, spark sources, and

Table 4.5 A list of explosive and potentially explosive chemical families.[a] Not all compounds in these chemical families are necessarily explosive, so if you are working with a compound that falls under one of these chemical families, double-check the SDS to see if it is categorized there as an explosive or potentially explosive compound.

Explosive and potentially explosive chemical families	
Acetylene (acetylide) compounds	Metal oxohalogenates
Acyl hypohalites	Metal oxometallates
Alkyl nitrates	Metal perchlorates
Alkyl perchlorates	Metal peroxides
Allyl trifluoromethanesulfonates	Metal peroxomolybdates
Amminemetal oxosalts	Metal picramates
Aromatic nitrates	Nitroaryl Compounds
Azides	*Aci*-Nitroquinonoid Compounds
Azocarbaboranes	*Aci*-nitro salts
N-Azolium nitroimidates	Nitroso compounds
Diazo compounds	N−S Compounds
Diazonium compounds	Organic acids
Difluoroaminoalkanols	Organolithium reagents
Fluoro−nitro compounds	Organomineral peroxides
Furazan N-oxides	Oximes
Hydroxooxodiperoxochromate salts	Oxosalts of nitrogenous bases
Iodine compounds	Ozonides
Isoxazoles	Perchlorate salts of nitrogenous bases
Metal azide halides	Perchloryl compounds
Metal azides	Peroxy compounds
N-Metal derivatives	Phosphorus esters
Metal fulminates	Picrates
Metal halogentates	Polynitroalkyl/aryl compounds
Metal hydrides	Strained-ring compounds
Metal nitrophenoxides	Tetrazoles
Metal oxides	Triazoles

[a] Extracted from UC Berkeley's *Guidelines for Explosive and Potentially Explosive Chemicals Safe Storage and Handling.* https://ehs.berkeley.edu/sites/default/files/pecguidelines.pdf. *Accessed October 27th, 2021.*

direct sunlight. In the event of a fire, you should also make sure to know what type of fire extinguishers are permissible to be used on these compounds, and that you know of the location of the closest fire extinguisher. This is especially important for flammable metals which require special fire extinguishers. In those cases, it might be good to keep the fire extinguisher close at hand, just in case something goes wrong. If the fire is large however or if there are other flammables close by that might ignite and/or cause an explosion, prioritize getting yourself to safety, warning others about the hazard, and then let the professionals take care of extinguishing the fire, informing them of what compounds caused the fire and what other hazardous chemicals are in the lab that might affect their safety.

If possible, wear a flame-resistant/retardant lab coat when working with these compounds, to reduce the risk of your lab coat going up in flames. Then when working with the flammables, do not use more than needed, and store the remaining flammable chemicals safely away. Flammable compounds are often volatile, so containers of flammable compounds that are not in use should be closed to prevent the vapors from reaching a heat source close by which might ignite it. Good general lab housekeeping practices can help minimize additional hazards from flammable chemicals.

Oxidizers can increase the intensity of fires and increase the combustibility/flammability of other chemicals, so make sure to practice good housekeeping when using and storing these compounds. They should not be stored alongside flammables, combustibles, and reducing agents as that could increase the risk of a fire. There are also many organic materials that would be safer to keep them away from. In some cases, reactions using oxidizers may also yield toxic gases, so remember to consider carefully potential side reactions of the experiments you perform, consulting the SDS and your coworkers before running new experiments.

Cryogenics and other extremely cold compounds

Liquids that have a boiling point below $-150°C$ are generally considered cryogenic. These are chemicals such as liquid nitrogen (N_2), helium (He), and argon (Ar). But scientists often work with other extremely cold compounds, like dry ice (CO_2, $T_{subl} = -78.5°C$), and cold solutions, such as those cooled down to $-78.5°C$ using dry ice, or solutions of organic solvents that have been brought down to their freezing point, e.g., using liquid nitrogen to form a slush of the organic solvent. When working with these low temperature liquids/solids/materials, it is important to be careful as they can cause severe frostbites if they come in contact with skin or other parts of the body. Proper gloves should be used and safety glasses when working with these chemicals (see Chapter 2 on gloves). When the cryogenic liquids escape into the atmosphere, they can evaporate quickly. As a small amount of liquid will form a large volume of gas this can create a different type of hazard. That is, in small rooms, or when transporting these liquids in an elevator, there is the danger of these cryogenics pushing the oxygen out of the room very quickly creating a different and hazardous situation. It is therefore important to be careful when working with cryogenics and other extremely cold compounds and be mindful of the potential risks so that you are better able to prevent an accident and/or react in a proper way to minimize any potential damage.

Working with extremely hazardous chemicals

If the plan is to work with extremely hazardous chemicals, such as carcinogens, or compounds that can cause mutations or birth defects, it is important to prepare yourself and the laboratory well for that type of work. Notify your coworkers what you are planning and limit access to the working area if needed. All toxic compounds should be handled in appropriate work areas such as fume hoods.

All use of hazardous chemicals is dependent on the worker having received adequate training. Supervisors need to make sure to provide the proper training for their team members/students, but you should also make sure to ask for the proper training. Some chemicals may also be so deadly that it is simply not worth working with them, at least not if you can avoid it. When dealing with these extremely hazardous chemicals, consider whether there is a different and safer way to do the research or experiments that you need to do, consulting your coworkers and/or supervisor in the process. If there is a safer way, use that, and avoid putting yourself in unnecessary danger. The case described here below shows the hazard of dimethylmercury, but as a result of this case, many chemists will not work with dimethylmercury any more or expose themselves to it in any way. Some experiments are just not worth the risk.

Deadly drops of dimethylmercury[3]

Back in 1996, a professor accidentally spilled several drops of dimethylmercury on her gloved hand. She was wearing all the proper PPE, including disposable latex gloves. After the spill, she cleaned it up and removed her gloves.

A few months later, the professor started to develop neurological problems. She went to the hospital and after detailed examination they found that she had methylmercury poisoning. This spill of only a few drops was the only exposure to mercury she had had and analysis of the mercury contents of her hair supported the timeline. Unfortunately, these gloves were not enough here and although treatments were attempted, she eventually passed away, 298 days after exposure. It is unsure if other types of gloves would have provided better protection, but given the super toxicity of dimethylmercury, it is recommended to avoid working with it as much as possible. If it becomes necessary to work with this compound, it should be handled with extreme caution.

[3] This case was extracted from: Nierenberg D.W. Delayed cerebellar disease and death after accidental exposure to dimethylmercury. *New Engl. J. Med.* 1998, *338*, 1672–1676.

Permanent and temporary storage containers

In general, you should always store the chemicals in their original containers. When it comes to temporary storage units for chemicals, e.g., during an experiment, consider the compatibility of the chemical with the material of the storage container. Some chemicals, like hydrofluoric acid for example, are not compatible with glass containers, and might react with it and leak through. In other cases, the compatibility might even lead to an explosive reaction, especially if it is closed. Always be careful about the choice of material in the chemical storage units. Tables 4.6 and 4.7 shows the general compatibilities of several types of containers and tubings, respectively.

Storage and labeling of synthesized products

Make sure to consider potential compatibility issues between your newly synthesized compounds and the storage units you decide to use. Table 4.6

Table 4.6 The compatibility of common compounds and storage materials and labware. *E*, Excellent; *F*, Fair; *G*, Good; *NR*, Not recommended.

Compounds	Labware											
	FEP	FLPE	HDPE	LDPE	PC	PSF	PS	PMMA	PETG	PMP	PP	PPCO
Acetaldehyde	E	E	G	G	NR	NR	NR	NR	–	G	G	G
Acetic acid	E	E	G	G	NR	G	F	NR	NR	E	E	E
Acetone	E	E	F	F	NR	F	NR	NR	NR	E	F	NR
Acetonitrile	E	E	E	E	NR	NR	NR	NR	–	F	E	F
Aluminum chloride	E	E	E	E	E	E	E	E	E	E	E	E
Ammonia (aqueous solution)	E	F	E	E	NR	G	E	NR	–	E	E	E
Ammonium hydroxyde	F	G	F	F	NR	G	G	F	NR	G	F	F
Benzyl alcohol	E	E	F	G	NR	G	NR	F	G	–	G	NR
Butanol	E	E	E	E	E	E	G	F	E	E	E	E
Butyl ether	E	E	–	NR	NR	NR	NR	NR	–	–	G	–
Calcium hydroxide	E	F	E	E	F	E	E	G	–	E	E	E
Chloroacetic acid	E	E	E	E	F	F	G	NR	–	E	E	E
Chloroform	E	G	F	F	NR	NR	NR	NR	NR	F	F	E
Chromic acid (80%)	E	NR	NR	NR	NR	NR	F	NR	NR	NR	NR	NR
Citric acid	E	E	E	G	E	E	E	E	–	E	E	E
Corn oil	E	E	E	G	E	E	E	E	G	E	E	E
Cyclohexanol	E	E	E	F	F	E	E	NR	–	E	E	E
Distilled water	E	E	E	E	E	E	E	E	E	E	E	E
Diethanolamine	E	G	–	–	F	F	E	–	E	–	E	–
Diethyl ether	E	G	NR	F	NR	NR	F	NR	E	NR	F	NR
Dimethyl acetamide	E	G	E	F	NR	NR	NR	–	–	G	E	E
Dimethylformamide	E	G	E	E	NR	NR	NR	NR	NR	E	E	E
Dimethylsulfoxide	E	E	E	NR	NR	NR	NR	NR	NR	E	E	E

Chemical												
Dioxane	E	E	G	G	NR	NR	NR	NR	—	F	F	G
Ethanol	E	E	E	E	G	E	G	NR	F	E	F	E
Ethanolamine	E	G	G	G	NR	NR	—	—	—	—	E	—
Ethyl acetate	E	E	E	E	NR	NR	NR	NR	NR	E	E	G
Ethylene glycol	E	E	E	E	E	E	E	G	E	E	E	E
Fatty acids	E	E	E	G	G	G	E	E	G	E	E	E
Formaline (10%)	E	E	E	E	E	E	G	E	—	E	E	E
Formaline (30%)	E	E	E	E	E	E	G	E	F	E	E	E
Formaline	E	E	E	E	F	E	G	NR	—	E	E	E
Formic acid	E	E	E	G	E	G	NR	E	—	E	E	E
Glycerole	E	E	E	E	F	E	E	E	E	E	E	E
Heptane	E	G	F	NR	E	F	E	G	G	F	F	—
Hydrochloric acid 1N	E	E	E	E	F	F	F	NR	F	E	E	E
Hydrochloric acid	E	E	E	E	NR	E	E	NR	NR	E	E	E
Hydrofluoric acid	E	E	E	E	E	E	G	E	G	E	E	E
Hydrogen peroxide (30%)	E	E	E	E	G	E	G	NR	—	E	E	E
Iodine (10%)	E	G	G	G	E	E	E	NR	F	E	E	E
Iso-butanol	E	E	E	E	G	E	E	E	F	E	E	E
Isopropyl alcohol	E	E	E	E	E	E	E	G	NR	E	E	E
Lactic acid	E	E	E	E	E	F	F	E	—	E	E	E
Maleic acid	E	E	E	E	NR	NR	NR	NR	—	—	E	—
Mercury	E	E	E	E	G	E	E	NR	NR	E	E	E
Methanol	E	E	F	G	E	E	E	E	G	E	E	E
Methyl ethyl ketone	E	E	E	G	E	E	F	NR	G	F	G	E
Mineral oil	E	E	E	G	E	E	NR	E	G	G	E	E
Naphthalene	E	G	F	NR	NR	NR	NR	NR	—	E	E	F

(Continued)

Table 4.6 The compatibility of common compounds and storage materials and labware. *E*, Excellent; *F*, Fair; *G*, Good; *NR*, Not recommended.—cont'd

Compounds	Labware											
	FEP	FLPE	HDPE	LDPE	PC	PSF	PS	PMMA	PETG	PMP	PP	PPCO
Nitric acid	E	F	F	F	F	F	NR	NR	NR	F	NR	NR
Oleic acid	E	G	G	F	E	E	F	E	NR	—	E	E
Oxalic acid	E	E	E	G	E	E	E	E	—	E	E	E
Pentane	E	G	NR	—	NR	NR	NR	E	—	—	G	—
Perchloric acid	E	E	E	E	NR	NR	G	F	—	—	E	—
Petroleum ether	E	F	F	NR	NR	NR	NR	G	E	NR	NR	NR
Phenol	E	E	F	F	NR	F	NR	NR	NR	NR	NR	G
Phosphoric acid	E	E	E	E	E	E	E	NR	—	E	E	E
Polyethylene glycol	E	E	E	E	E	E	E	—	—	—	E	—
Potassium chloride	E	E	E	E	E	E	E	E	E	E	E	E
Potassium hydroxide	E	F	E	E	NR	E	E	E	NR	E	E	E
Potassium permanganate	E	E	G	E	E	E	E	G	—	E	E	E
Propanol	E	E	E	E	E	E	E	NR	F	E	E	E
Silicon oil	E	E	E	E	E	E	E	F	G	E	E	E
Silver nitrate (10%)	E	E	E	E	E	E	E	E	F	E	E	E
Sodium chloride	E	E	E	E	E	E	E	E	—	E	E	E
Sodium hydroxide	E	E	E	E	NR	F	E	NR	NR	E	E	E
Sodium lauryl sulfate	E	E	E	E	E	E	E	E	G	E	E	E
Sucrose	E	E	E	E	E	E	E	E	E	E	E	E
Sulfuric acid (10%)	E	E	E	E	E	E	F	E	E	E	E	E
Sulfuric acid	E	F	NR	G	NR	NR	NR	NR	NR	NR	F	NR
Tetrahydrofuran	E	G	F	F	NR	NR	NR	NR	—	F	G	G
Toluene	E	F	NR	NR	NR	NR	NR	NR	NR	F	NR	NR

Trichloroacetic acid	E	F	F	E	E	F	G	FF	NR	—	E	G	F
Triethanolamine	E	G	E	E	E	NR	NR	E	E	—	—	F	—
Tris buffer soltuion	E	E	E	E	G	G	G	E	E	E	E	E	E
Tween 20	E	E	E	G	F	F	G	E	E	E	E	E	E
Urea (10%)	E	E	E	E	E	E	E	E	E	—	E	E	E
Xylene	E	G	G	F	NR	NR	NR	NR	NR	—	F	F	F
Zinc sulfide	E	E	E	E	E	E	E	E	E	E	E	E	E

FEP, Fluorinated ethylene propylene; *FLPE*, Fluorinated high–density polyethylene; *HDPE*, High density polyethylene; *LDPE*, Low density polyethylene; *PC*, Polycarbonate; *PETG*, Polyethylene terephthalate co-polyester; *PMMA*, Polymethyl methacrylate; *PMP*, Polymethylpentene; *PP*, Polypropylene; *PPCO*, Polypropylene copolymer; *PS*, Polystyrene; *PSF*, Polysufone.

Table 4.7 The compatibility of common compounds and various tubing materials. *E*, Excellent; *F*, Fair; *G*, Good; *NR*, Not recommended.

Compounds	Tubings					
	PUR	PVC	PFA	FEP	Silicone	PP
Acetaldehyde	NR	NR	E	E	F	G
Acetic acid	NR	NR	E	E	NR	E
Acetone	NR	NR	E	E	NR	F
Acetonitrile	NR	NR	E	E	NR	E
Aluminum chloride	E	G	E	E	F	E
Ammonia (aqueous solution)	F	G	E	E	NR	E
Ammonium hydroxide	G	F	F	F	E	F
Benzyl alcohol	NR	F	E	E	F	G
Butanol	NR	G	E	E	F	E
Butyl ether	–	–	E	E	NR	G
Calcium hydroxide	E	E	E	E	E	E
Chloroacetic acid	NR	NR	E	E	NR	E
Chloroform	NR	NR	E	E	NR	NR
Chromic acid (80%)	NR	NR	E	E	NR	F
Citric acid	G	F	E	E	E	E
Corn oil	–	NR	E	E	NR	E
Cyclohexanol	–	NR	E	E	NR	E
Distilled water	E	E	E	E	E	E
Diethanolamine	–	–	E	E	F	E
Diethyl ether	NR	NR	E	E	NR	NR
Dimethyl acetamide	NR	NR	E	E	NR	E
Dimethylformamide	NR	NR	E	E	F	E
Dimethylsulfoxide	NR	NR	E	E	NR	E
Dioxane	NR	NR	E	E	NR	F
Ethanol	NR	F	E	E	F	E
Ethanolamine	–	–	E	E	F	E
Ethyl acetate	NR	NR	E	E	F	E
Ethylene glycol	E	F	E	E	E	E
Fatty acids	G	E	E	E	F	E
Formaline (10%)	–	G	E	E	F	E
Formaline (30%)	NR	F	E	E	F	E
Formaline	NR	F	E	E	F	E
Formic acid	NR	F	E	E	F	E
Glycerole	E	F	E	E	E	E
Heptane	E	NR	E	E	NR	F
Hydrochloric acid 1N	F	G	E	E	E	E
Hydrochloric acid	NR	F	E	E	NR	E
Hydrofluoric acid	NR	F	E	E	NR	E
Hydrogen peroxide (30%)	F	G	E	E	E	E
Iodine (10%)	NR	NR	E	E	F	E
Iso-butanol	NR	G	E	E	E	E
Isopropyl alcohol	NR	G	E	E	NR	E

Table 4.7 The compatibility of common compounds and various tubing materials. *E*, Excellent; *F*, Fair; *G*, Good; *NR*, Not recommended.—cont'd

Compounds	Tubings					
	PUR	PVC	PFA	FEP	Silicone	PP
Lactic acid	NR	G	E	E	G	E
Maleic acid	G	E	E	E	F	E
Mercury	E	G	E	E	NR	E
Methanol	NR	F	E	E	E	E
Methyl ethyl ketone	NR	NR	E	E	NR	G
Mineral oil	E	F	E	E	E	E
Naphthalene	G	NR	E	E	NR	E
Nitric acid	NR	NR	E	E	NR	NR
Oleic acid	G	NR	E	E	NR	E
Oxalic acid	NR	F	E	E	F	E
Pentane	E	NR	E	E	NR	G
Perchloric acid	NR	NR	F	E	NR	E
Petroleum ether	NR	NR	E	E	NR	NR
Phenol	NR	F	E	E	NR	NR
Phosphoric acid	NR	G	E	E	NR	E
Polyethylene glycol	G	G	E	E	NR	E
Potassium chloride	E	E	E	E	E	E
Potassium hydroxide	NR	F	E	E	NR	E
Potassium permanganate	E	G	E	E	E	E
Propanol	NR	G	E	E	E	E
Silicon oil	E	G	E	E	NR	E
Silver nitrate (10%)	E	E	E	E	E	E
Sodium chloride	E	E	E	E	E	E
Sodum hydroxide	NR	F	E	E	NR	E
Sodium lauryl sulfate	G	G	E	E	E	E
Sucrose	E	E	E	E	E	E
Sulfuric acid (10%)	G	E	E	E	NR	E
Sulfuric acid	NR	NR	E	E	NR	F
Tetrahydrofuran	NR	NR	E	E	NR	G
Toluene	NR	NR	E	E	NR	NR
Trichloroacetic acid	—	G	E	E	NR	G
Triethanolamine	NR	—	E	E	F	F
Tris buffer solution	G	E	E	E	E	E
Tween 20	G	F	E	E	E	E
Urea (10%)	G	G	E	E	E	E
Xylene	NR	NR	E	E	NR	F
Zinc sulfide	E	G	E	E	E	E

FEP, Fluorinated ethylene propylene; *PFA*, Perfluoroalkoxy alkanes; *PP*, Polypropylene; *PUR*, Polyurethane; *PVC*, Polyvinyl chloride.

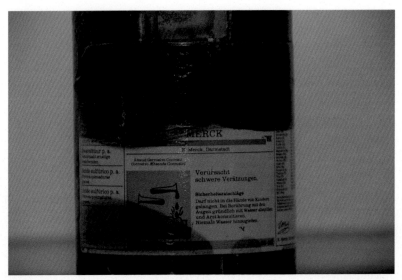

Figure 4.4 Be careful that your labels are always clear and do not get smudged like the one shown here. If that happens, update the label right away while you still remember what compound/product the container contains. *(Credit: Sveinbjörn Gizurarson).*

here below may be of use there. It is also crucial to label all synthesized compounds clearly. No unlabeled chemicals should be found in your work area. Therefore, make it a practice to label the chemicals clearly and right away, and be careful that the labels do not get smudged or washed away (Fig. 4.4). If that happens, relabel them again right away while you still remember what compound they contain.

Hazard and precautionary statements (H- and P-codes)

The hazard and precautionary statements are regularly updated. In this list, we include a combination of the last few updates of hazard and precautionary statements (Table 4.8), with those marked * having been deleted in recent GHS revisions, and not being present in the 2021 revision. These are included as they are likely still in circulation on many older chemical containers that might still be in use, especially in research laboratories.

Table 4.8 Hazard statements (H) and precautionary statements (P) from globally harmonized system of classification and labelling of chemicals (GHS).

Physical hazards

H200*: Unstable explosive
H201*: Explosive; mass explosion hazard
H202*: Explosive; severe projection hazard
H203*: Explosive; fire, blast or projection hazard
H204: Fire or projection hazard
H205*: May mass explode in fire
H206: Fire, blast or projection hazard; increased risk of explosion if desensitizing agent is reduced
H207: Fire or projection hazard; increased risk of explosion if desensitizing agent is reduced
H208: Fire hazard; increased risk of explosion if desensitizing agent is reduced
H209: Explosive
H210: Very sensitive
H211: May be sensitive
H220: Extremely flammable gas
H221: Flammable gas
H222: Extremely flammable aerosol
H223: Flammable aerosol
H224: Extremely flammable liquid and vapour
H225: Highly flammable liquid and vapour
H226: Flammable liquid and vapour
H227: Combustible liquid
H228: Flammable solid
H229: Pressurized container: may burst if heated
H230: May react explosively even in the absence of air
H231: May react explosively even in the absence of air at elevated pressure and/or temperature
H232: May ignite spontaneously if exposed to air
H240: Heating may cause an explosion
H241: Heating may cause a fire or explosion
H242: Heating may cause a fire
H250: Catches fire spontaneously if exposed to air
H251: Self-heating; may catch fire
H252: Self-heating in large quantities; may catch fire
H260: In contact with water releases flammable gases which may ignite spontaneously
H261: In contact with water releases flammable gas

(Continued)

Table 4.8 Hazard statements (H) and precautionary statements (P) from globally harmonized system of classification and labelling of chemicals (GHS).—cont'd

H270: May cause or intensify fire; oxidizer
H271: May cause fire or explosion; strong oxidizer
H272: May intensify fire; oxidizer
H280: Contains gas under pressure; may explode if heated
H281: Contains refrigerated gas; may cause cryogenic burns or injury
H282: Extremely flammable chemical under pressure: may explode if heated
H283: Flammable chemical under pressure: may explode if heated
H284: Chemical under pressure: may explode if heated
H290: May be corrosive to metals

Health hazards

H300: Fatal if swallowed
H301: Toxic if swallowed
H302: Harmful if swallowed
H303: May be harmful if swallowed
H304: May be fatal if swallowed and enters airways
H305: May be harmful if swallowed and enters airways
H310: Fatal in contact with skin
H311: Toxic in contact with skin
H312: Harmful in contact with skin
H313: May be harmful in contact with skin
H314: Causes severe skin burns and eye damage
H315: Causes skin irritation
H316: Causes mild skin irritation
H317: May cause an allergic skin reaction
H318: Causes serious eye damage
H319: Causes serious eye irritation
H320: Causes eye irritation
H330: Fatal if inhaled
H331: Toxic if inhaled
H332: Harmful if inhaled
H333: May be harmful if inhaled
H334: May cause allergy or asthma symptoms or breathing difficulties if
      inhaled
H335: May cause respiratory irritation
H336: May cause drowsiness or dizziness
H340: May cause genetic defects
H341: Suspected of causing genetic defects
H350: May cause cancer

(Continued)

Table 4.8 Hazard statements (H) and precautionary statements (P) from globally harmonized system of classification and labelling of chemicals (GHS).—cont'd

H350i: May cause cancer by inhalation
H351: Suspected of causing cancer
H360: May damage fertility or the unborn child
H360F*: May damage fertility
H360D*: May damage the unborn child
H360FD*: May damage fertility; May damage the unborn child
H360Fd*: May damage fertility; Suspected of damaging the unborn child
H360Df*: May damage the unborn child; Suspected of damaging fertility
H361: Suspected of damaging fertility or the unborn child
H361f*: Suspected of damaging fertility
H361d*: Suspected of damaging the unborn child
H361fd*: Suspected of damaging fertility; Suspected of damaging the unborn child
H362: May cause harm to breast-fed children
H370: Causes damage to organs
H371: May cause damage to organs
H372: Causes damage to organs through prolonged or repeated exposure
H373: May cause damage to organs through prolonged or repeated exposure

Environmental hazards

H400: Very toxic to aquatic life
H401: Toxic to aquatic life
H402: Harmful to aquatic life
H410: Very toxic to aquatic life with long-lasting effects
H411: Toxic to aquatic life with long-lasting effects
H412: Harmful to aquatic life with long-lasting effects
H413: May cause long-lasting harmful effects to aquatic life
H420: Harms public health and the environment by destroying ozone in the upper atmosphere

General precautionary statements

P101: If medical advice is needed, have product container or label at hand.
P102: Keep out of reach of children
P103: Read label before use

Prevention precautionary statements

P201*: Obtain special instructions before use
P202*: Do not handle until all safety precautions have been read and understood

(Continued)

Table 4.8 Hazard statements (H) and precautionary statements (P) from globally harmonized system of classification and labelling of chemicals (GHS).—cont'd

P203: Obtain, read and follow all safety instructions before use

P210: Keep away from heat, hot surfaces, sparks, open flames and other ignition sources. No smoking

P211: Do not spray on an open flame or other ignition source

P212: Avoid heating under confinement or reduction of the desensitized agent

P220: Keep away from clothing and other combustible materials

P221*: Take any precaution to avoid mixing with combustibles/…

P222: Do not allow contact with air

P223: Do not allow contact with water

P230: Keep wetted with …

P231: Handle under inert gas

P232: Protect from moisture

P233: Keep container tightly closed

P234: Keep only in original container

P235: Keep cool

P236: Keep only in original packaging; Division … in the transport configuration

P240: Ground/bond container and receiving equipment

P241: Use explosion-proof [electrical/ventilating/lighting/…/] equipment

P242: Use only non-sparking tools

P243: Take precautionary measures against static discharge

P244: Keep valves and fittings free from oil and grease

P250: Do not subject to grinding/shock/friction/…

P251: Do not pierce or burn, even after use

P260: Do not breathe dust/fume/gas/mist/vapors/spray

P261: Avoid breathing dust/fumes/gas/mist/vapors/spray

P262: Do not get in eyes, on skin, or on clothing

P263: Avoid contact during pregnancy/while nursing.

P264: Wash … thoroughly after handling

P265: Do not touch eyes

P270: Do not eat, drink or smoke when using this product

P271: Use only outdoors or in a well-ventilated area

P272: Contaminated work clothing should not be allowed out of the workplace

P273: Avoid release to the environment

P280: Wear protective gloves/protective clothing/eye protection/face protection

P281*: Use personal protective equipment as required

P282: Wear cold insulating gloves/face shield/eye protection

(Continued)

Table 4.8 Hazard statements (H) and precautionary statements (P) from globally harmonized system of classification and labelling of chemicals (GHS).—cont'd

P283: Wear fire resistant or flame retardant clothing
P284: [In case of inadequate ventilation] wear respiratory protection
P285*: In case of inadequate ventilation wear respiratory protection

Response precautionary statements

P301: IF SWALLOWED:
P302: IF ON SKIN:
P303: IF ON SKIN (or hair):
P304: IF INHALED:
P305: IF IN EYES:
P306: IF ON CLOTHING:
P307: IF exposed:
P308: IF exposed or concerned:
P309*: IF exposed or if you feel unwell
P310*: Immediately call a POISON CENTER or doctor/physician
P311*: Call a POISON CENTER or doctor/...
P312*: Call a POISON CENTER or doctor/... if you feel unwell
P313*: Get medical advice/attention
P314*: Get medical advice/attention if you feel unwell
P315*: Get immediate medical advice/attention
P316: Get emergency medical help immediately
P317: Get medical help
P318: IF exposed or concerned, get medical advice
P319: Get medical help if you feel unwell
P320: Specific treatment is urgent (see ... on this label)
P321: Specific treatment (see ... on this label)
P322*: Specific measures (see ... on this label)
P330: Rinse mouth
P331: Do NOT induce vomiting
P332: If skin irritation occurs:
P333: If skin irritation or a rash occurs:
P334: Immerse in cool water [or wrap in wet bandages]
P335: Brush off loose particles from skin.
P336: Thaw frosted parts with lukewarm water. Do not rub affected areas
P337: If eye irritation persists:
P338: Remove contact lenses, if present and easy to do. Continue rinsing
P340: Remove victim to fresh air and keep at rest in a position comfortable
        for breathing

(Continued)

Table 4.8 Hazard statements (H) and precautionary statements (P) from globally harmonized system of classification and labelling of chemicals (GHS).—cont'd

P341*: If breathing is difficult, remove victim to fresh air and keep at rest in a position comfortable for breathing

P342: If experiencing respiratory symptoms:

P350*: Gently wash with plenty of soap and water

P351: Rinse cautiously with water for several minutes

P352: Wash with plenty of water/...

P353: Rinse skin with water [or shower]

P354: Immediately rinse with water for several minutes

P360: Rinse immediately contaminated clothing and skin with plenty of water before removing clothes

P361: Take off immediately all contaminated clothing

P362: Take off contaminated clothing

P363: Wash contaminated clothing before reuse

P364: And wash it before reuse

P370: In case of fire:

P371: In case of major fire and large quantities:

P372: Explosion risk

P373: DO NOT fight fire when fire reaches explosives

P374*: Fight fire with normal precautions from a reasonable distance

P375: Fight fire remotely due to the risk of explosion

P376: Stop leak if safe to do so

P377: Leaking gas fire: Do not extinguish, unless leak can be stopped safely

P378: Use ... to extinguish

P380: Evacuate area

P381: In case of leakage, eliminate all ignition sources

P390: Absorb spillage to prevent material damage

P391: Collect spillage.

Storage precautionary statements

P401: Store in accordance with ...

P402: Store in a dry place

P403: Store in a well ventilated place

P404: Store in a closed container

P405: Store locked up

P406: Store in a corrosive resistant/... container with a resistant inner liner.

P407: Maintain air gap between stacks or pallets

P410: Protect from sunlight

P411: Store at temperatures not exceeding ... °C/... °F.

P412: Do not expose to temperatures exceeding 50 °C/122 °F.

P413: Store bulk masses greater than ... kg/... lbs at temperatures not exceeding ... °C/...°F.

(Continued)

Table 4.8 Hazard statements (H) and precautionary statements (P) from globally harmonized system of classification and labelling of chemicals (GHS).—cont'd

P420: Store separately
P422*: Store contents under ...

Disposal precautionary statements

P501: Dispose of contents/container to ...
P502: Refer to manufacturer or supplier for information on recovery or recycling
P503: Refer to manufacturer/supplier... for information on disposal/recover/recycling

D, effect on unborn child; *F*, effect on fertilty.
(Published here with the permission of the UN.).

CHAPTER 5

Biohazards

Introduction

Laboratories, where people work with biohazards, need to fulfill strict requirements for safety. We do not want any microbes to escape into the environment, possibly spreading a disease. This includes laboratories where students, scientists, or researchers are working with microbes such as genetically mutated organisms or biological samples. Most countries have special law and regulations for these laboratories, that other labs do not require. Since all samples in these laboratories are biological and may cause infection and harm to the people working there, these facilities should be strictly monitored and there needs to be limited access to these laboratories. If something goes wrong, the results can be consequential, and even stretch outwards to the society. The following are considered as potential bio-hazards: microbes such as bacteria, viruses, parasites, prions, and fungi (both wild strains and/or genetically modified), all biological material, especially human samples, as well as allergens and toxins. Under this category we also have genetic material (DNA and RNA) and materials processed from biological materials such as plants, animals, or humans.

All biological samples, whether isolated microbes, a blood sample, swab, or a tissue sample, should be handled with utmost care. Treat all samples as if they were contaminated with an extremely contagious virus. Samples like blood, amniotic fluid, synovial fluid, spinal fluid, cervical mucus, semen, and fluids surrounding the heart and other organs are considered samples of body fluids and should also be treated as such. Other biological samples like saliva (spit), stool, urine, vomit, tear, sweat, and mucus from the respiratory system should also be treated with utmost care, as if they contained an extremely contagious virus.

There are many ways transmission can occur within the laboratory. When handling the sample, small droplets may get into the air, mist, or micro- or nanoparticles could get aerosolized and be dispersed into the environment. Whenever infections take place, it is most often related to

Handbook for Laboratory Safety
ISBN 978-0-323-99320-3
https://doi.org/10.1016/B978-0-323-99320-3.00003-3

accidents and/or moments of carelessness or negligence. Some infections are due to inhalation of mist or aerosolized droplets that can be formed during the mixing of samples, when a stopper is removed from a bottle or when samples are removed from a centrifuge. This mist, or aerosolized droplets, can end up on, or nearby, surface areas that someone then touches, resulting in an infected researcher. Other transmission pathways can relate to accidents like needle puncture, animal bites (e.g., in field work), scratches from poor glassware, or accidents connected to sample spills. It is therefore important to wear gloves, a lab coat, and a mask, all the time, when working in these working environments.

Good work ethics, following proper protocols, and vigilance are the most important ways to prevent infections. The United States Center for Disease Control (CDC) estimates that every year there are about 380,000 puncture wounds in laboratories, but not all of these accidents lead to infections. The most common infection pathways in lab workers are due to accidents like needle puncture, or when samples get spilled on the skin, or if drops or mist end up on the mucous membrane (nose, mouth, or eye). Note that aerosolized drops may be formed when stoppers pop of bottles or when removed from tubes.

Infected undergraduate students[1]

Two 20-year-old students, from the Australian Capital Territory, came to the emergency department with a three-day history of vomiting, abdominal cramps, and watery diarrhea. They were admitted to the hospital and treated with intravenous fluids and oral antibiotics. Stool samples were positive for *Salmonella enterica* and *Salmonella typhimurium*. Both students had participated in an undergraduate microbiological course at a university in Canberra, two weeks prior to symptom onset. Both patients recovered and were discharged after treatment.

Always have your mind focused on your work, when working with potential hazards.

[1] This case was extracted from: Sloan-Gardner, T.S.; Experimenting at Uni: *Salmonella* in laboratory students. *Commun. Dis. Intell.* 2018, *43*, https://doi.org/10.33321/cdi.2019.43.38.

Hand washing

Gloves and hand washing are the most important personal protections in the laboratory. It is a good practice to disinfect your gloves, followed by washing your hands when you leave the laboratory, even though you wore gloves the whole time while you were working there. Putting up the gloves should be the first thing you do upon entering the laboratory and never take them off. For certain labs, it may even be better to put the gloves on right before entering the laboratory. Taking the gloves off should be one of the last things you do when you leave the lab. At that time, take off your gloves and wash your hands before you touch anything else, especially before touching your face, your clothes, going to the bathroom, eating, drinking, or anything else.

Careful handwashing includes the use of soap (Fig. 5.1). Liquid soap and water significantly reduce the transmission probability from potential biohazards that might have remained on your hands, or been transmitted to your hands while taking off the gloves (Fig. 5.2). When working with especially dangerous biohazards, higher standards need to be followed, which often includes the use of disinfectants or antiseptics to secure the removal of possible microbes on your lab coat and gloves. When working with harmful microbes such as the SARS-CoV-2 virus, the researcher must wear a full protection, to protect all possible skin and mucosal surfaces from being exposed to the microbe.

Laboratory-acquired dengue virus[2]

In August 2018 a laboratory worker was admitted with dengue. He had not been traveling, so his laboratory was visited looking at safety protocols and other requirements. The patient had been wearing a single pair of nitrile gloves, eye protection, a lab coat, and closed-toe shoes while working with an infectious virus in a certified biosafety cabinet. When interviewing the patient, he reported that small splashes often occurred during virus production and purification. The patient did not change gloves when splashes occurred but occasionally performed surface decontamination of gloves. He had not been vigilant about handwashing after removing gloves.

Additionally, the patient had a compression wound on the ring finger of the left hand, that was infected and oozing. He did not cover the wound before donning a single pair of gloves. When the patient was asked to demonstrate how he removed his gloves, it showed that he could have contaminated his finger, then potentially touched a mucosal surface of the nose or mouth. Or he could have touched a mucosal surface with the lab coat sleeve.

[2] This case was extracted from: Sharp, T.M.; Fisher, T.G.; Long, K.; Coulson, G.; Medina, F.A.; Herzig, C.; Koza, M.B.; Muñoz-Jordán, J.; Paz-Bailey, G.; Moore, Z.; Williams, C.; Laboratory acquired dengue virus infection, United States, 2018. *Emerg. Infect. Dis.* 2020, *26* (7), 1534–1537.

How to Handwash?

WASH HANDS WHEN VISIBLY SOILED! OTHERWISE, USE HANDRUB

🕐 **Duration of the entire procedure:** 40-60 seconds

Wet hands with water;

Apply enough soap to cover
all hand surfaces;

Rub hands palm to palm;

Right palm over left dorsum with
interlaced fingers and vice versa;

Palm to palm with fingers interlaced;

Backs of fingers to opposing palms
with fingers interlocked;

Rotational rubbing of left thumb
clasped in right palm and vice versa;

Rotational rubbing, backwards and
forwards with clasped fingers of right
hand in left palm and vice versa;

Rinse hands with water;

Dry hands thoroughly
with a single use towel;

Use towel to turn off faucet;

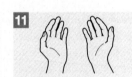

Your hands are now safe.

Figure 5.1 World Health Organization guidelines for proper hand washing. *(Credit: WHO.)*

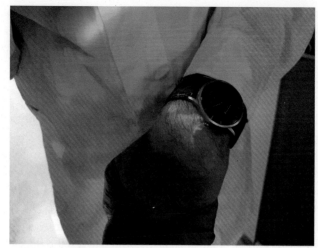

Figure 5.2 Luminated areas show contaminated skin, after removing the gloves. *(Credit: Benjamín Ragnar Sveinbjörnsson.)*

Exercise

It is good to train people in removing their gloves without being contaminated. This can be done by using a fluid/powder that glows under UV-light and ask them to rub the fluid all over their gloves, and then remove the gloves. Then turn on a UV-light to check if any part of their hands was contaminated with the UV–visible fluid/powder. Common areas that get contaminated include the fingers and right below the palm (Fig. 5.2).

Biological safety laboratory

Before work starts and again when it is finished, you should disinfect the work area. That should be done at the start for the integrity of the samples that you will be working with and at the end to keep the working area clean and to reduce the odds of infection. All tools (scissors, tweezers (pincers), etc.) that are used in these laboratories should be sterilized before use. After use, they should be sanitized, cleaned, and sterilized again. Instruments such as centrifuges should be cleaned regularly with a disinfectant to ensure that no contamination take place between samples but also to protect staff against infection. The following protocol should be followed in these labs:

- All tools that get in touch with biohazards should be disinfected and sterilized.
- All cabinets, refrigerators, and freezers that are used to store these samples should be well labeled as such and be disinfected regularly.
- Samples that are put into a centrifuge should be closed tightly and treated in a manner to minimize the risk of contamination from the sample.
- Sample vials with rubber stoppers should be opened carefully, so that droplets do not get transmitted to the person holding the vial.
- It is strictly forbidden to use the mouth to draw up a sample. Adequate pipettes should be available at all workplaces.
- Do not fix your hair, scratch your nose, touch your eyes, or anything else with the gloves. The outside of your gloves should never touch your skin.
- Do not put on lipstick or other make-up items inside the laboratory.
- You should know where the eye wash stations and emergency showers are located.

Hazard classes

Microbes are classified into hazard classes depending on the danger that they can cause to humans. Some microbes rarely or never cause infections, while others are life-threatening, even in microdroplet amounts. The various types and strains of viruses, bacteria, and fungi have been classified into four classes (see Table 5.1). Similarly, the laboratories that are allowed to work with these microbes are also classified into four biosafety levels (BSL-1, BSL-2, BSL-3, and BSL-4) (Fig. 5.3).

BSL-1 laboratories: Here it is permitted to work with microbes that are considered harmless or very unlikely to cause infections in humans. You need basic training to work in BSL-1 laboratories, as well as a hands-on training for BSL-1 work. General training in handling biohazards is recommended. It is also important that someone experienced in working in BSL-1 will provide guidance for all new students or employees when they start, to prevent any mistakes. The following guidelines should be used in BSL-1 laboratories:

- The work area should be smooth and easy to clean.
- A hand wash station should be present. Hands must be washed with soap and if you get contaminated, wash your hands immediately, and disinfect them.

Table 5.1 Microbes are classified depending on the severity of infection, risk of pandemic, and treatment options. Here is a list of bacteria, parasites, fungi, viruses, and prions that are in Class 3 and 4.

Bacteria	Class
Bacillus anthracis	3
Brucella abortus	3
Brucella canis	3
Brucella melitensis	3
Brucella suis	3
Burkholderia mallei (Pseudomonas mallei)	3
Burkholderia pseudomallei (Pseudomonas pseudomallei)	3
Chlamydia psittaci	3
Coxiella burnetii	3
Escherichia coli such as O157:H7, O103, etc.	3
Francisella tularensis (type A)	3
Mycobacterium africanum	3
Mycobacterium bovis (except BCG-strain)	3
Mycobacterium leprae	3
Mycobacterium microti	3
Mycobacterium tuberculosis	3
Mycobacterium ulcerans	3
Rickettsia akari	3
Rickettsia canada	3
Rickettsia conorii	3
Rickettsia montana	3
Rickettsia typhi (Rickettsia mooseri)	3
Rickettsia prowazekii	3
Rickettsia rickettsii	3
Rickettsia tsutsugamushi	3
Salmonella Typhi	3
Shigella dysenteriae (type 1)	3
Yersinia pestis	3
Parasite	**Class**
Echinococcus granulosus	3
Echinococcus multilocularis	3
Echinococcus vogeli	3
Leishmania brasiliensis	3
Leishmania donovani	3
Naegleria fowleri	3
Plasmodium falciparum (malaria)	3
Taenia solium	3

Continued

Table 5.1 Microbes are classified depending on the severity of infection, risk of pandemic, and treatment options. Here is a list of bacteria, parasites, fungi, viruses, and prions that are in Class 3 and 4.—cont'd

Parasite	Class
Trypanosoma brucei rhodesiense 3	3
Trypanosoma cruzi	3

Fungi	Class
Blastomyces dermatitidis (ajellomyces dermatitidis)	3
Cladophialophora bantiana (previously known as: *Xylohypha bantiana, Cladosporium bantianum,* or *trichoides)*	3
Coccidioides immitis	3
Histoplasma capsulatum var. capsulatum (ajellomyces capsulatus)	3
Histoplasma capsulatum duboisii	3
Paracoccidioides brasiliensis	3

Virus	Class
Arenaviridae	
Encephalitis viral	3
Flexal	3
Guanarito	4
Junin virus	4
Lassa fever	4
Machupo virus	4
Sabia	4
Bunyaviridae	
Belgrade (also known as Dobrava)	3
Oropouchevirus	3
Sin Nobre (previously known as Muerto Canyon)	3
Caliciviridae	
Hepatitis E	3
Filoviridae	
Ebola virus	4
Marburg virus	4
Flaviviridae	
Australian encephalitis (Murray Valley encephalitis)	3
Absettarov	3

Table 5.1 Microbes are classified depending on the severity of infection, risk of pandemic, and treatment options. Here is a list of bacteria, parasites, fungi, viruses, and prions that are in Class 3 and 4.—cont'd

Virus	Class
Dengue virus type 1—4	3
Hanzalova	3
Hepatitis C	3
Hepatitis G	3
Japanese encephalitis type B	3
Kumlinge	3
Kyasanur Forest	3
Louping ill virus	3
Omsk hemorrhagic fever virus	3
Powassan encephalitis virus	3
Rocio viral encephalitis	3
Russian spring-summer encephalitis virus	3
St. Louis encephalitis	3
Tick-born encephalitis (Central European encephalitis)	3
Wesselsbron virus	3
West-Nile virus	3
Yellow fever	3
Flebovirus	
Rift Valley disease	3
Hanta virus	
Hantaan (Korean hemorrhagic fever)	3
Seoulveira	3
Hepadnaviridae	
Hepatitis B	3
Hepatitis D	3
Herpesviridae	
Herpes simplex virus	3
Nairovirus	
Crimean-Congo hemorrhagic fever	4
Poxviridae	
Monkeypox virus	3
Orthopoxvirus	4
Parapoxvirus	4

Continued

Table 5.1 Microbes are classified depending on the severity of infection, risk of pandemic, and treatment options. Here is a list of bacteria, parasites, fungi, viruses, and prions that are in Class 3 and 4.—cont'd

Virus	Class
Smallpox virus (variola)	4
Vaccinia virus	4
Retroviridae	
Immunosuppressive viruses	3
T-cell viruses such as HTLV type 1 and 2	3
SIV virus	3
Rhabdoviridae	
Rabies virus	3
Togaviridae	
Eastern equine encephalitis virus (sleeping sickness)	3
Chikungunya virus	3
Everglades virus	3
Mayaro virus	3
Mucambo virus	3
Ndumu virus	3
Tonate virus	3
Venezuelan equine encephalitis virus	3
Western equine encephalitis virus	3
Toroviruses	
Human coronavirus (such as OC43, 229E, NL63, HKU1)	4
Severe acute respiratory syndrome coronavirus (SARS-CoV-2)	4
Unclassified viruses	
Hepatitis (unclassified)	3
Equine morbillivirus	4
Prions	
BSE (Bovine spongiform encephalopati)	3
Creutzfeldt—Jakob's disease	3
Gerstmann—Straeussler—Scheinker syndrome	3
Kuru	3
Variants of Creutzfeldt—Jacob's disease	3

Figure 5.3 The biohazard sign.

- Personal protective equipment is required, such as lab coat, gloves, and a mask. Never wear PPE used inside BSL-1, outside the laboratory.
- If you have a scratch or a wound, cover well with a bandage.
- Appropriate disinfectants should be easily accessible in the laboratory.
- The working area should be cleaned and disinfected after use.
- In case of contamination, the whole area should be disinfected.
- Glassware and other tools used should be washed, disinfected, and sterilized.
- Work should in general be conducted in a biological safety cabinet (BSC) Class I, to limit the dispersion of the sample.
- All waste should be bagged properly and disinfected, e.g., using an autoclave.
- All waste should be transmitted in secure, closed containers.

BSL-2 laboratories: In BSL-2 laboratories it is allowed to work with microbes that can cause infection in humans but are unlikely to result in an epidemic. Contact with infected samples rarely cause infection, and in case it happens there exists a well-established treatment or prevention. Examples of microbes in this class are salmonella, the measles virus, and toxoplasma. No one is permitted to work in these laboratories unless they have received BSL-2 training upon joining the laboratory, have received hands-on training for BSL-2 work, and annually review the laboratory safety plans and procedures, as well as go through a general training in handling of biohazards. In addition to the requirements listed for BSL-1, the following should be added:

- Access to the lab should be restricted to those who have received proper training and have the authorization to work there. All others, such as guests, are not permitted in the lab.

↙ A hand wash station should be at the exit of the lab. The faucets should be in such a manner that it is possible to control them without touching them directly.

↙ Autoclaves should be accessible to sterilize contaminated waste.

↙ Syringes, needles, or other sharp objects should not be used if it can be avoided. Those items should be stored in appropriate containers to prevent risk of cuts or puncture. Do not try to re-cover the needles after use but dispose of them in appropriate sharp waste collectors. Needles should never be bent or handled in a carefree manner with your hands.

↙ PPE intended for BSL-2 should be kept separately from other PPE. Ideally, lab coats should have long sleeves and be buttoned at the side. They should always be completely buttoned up.

↙ Work should be conducted in BSC Class II, to limit the dispersion of the sample.

BSL-3 laboratories: BSL-3 are designed to allow work with microbes that can cause serious infections in people. These microbes may also spread and could result in an epidemy, but treatments or preventative treatments are available. Examples of BSL-3 pathogens are tuberculosis (*Mycobacterium tuberculosis*), leprosy (*Mycobacterium leprae*), anthrax (*Bacillus anthracis*), Hepatitis B, and the AIDS virus (HIV). All contact with samples requires immediate response and treatment. No one is permitted to enter these laboratories unless they have received BSL-3 training, hands-on training for BSL-3 work, and review annually the laboratory safety plans and procedures, as well as advanced training in handling of biohazards. In addition to the requirements listed for BSL-1 and BSL-2, the following should be added:

↙ The laboratory must be designed in such a manner that it is possible to disinfect all surface areas, including walls, floor, and ceiling through fumigation. All windows should be closed and tight.

↙ The lab should be isolated from all traffic, if possible. The doors to the working area should always be securely locked and the only access should be through two doors, where you can change your clothes before entering the BSL-3 and when leaving the BSL-3.

↙ The entrance to the BSL-3 should be well labeled with the hazard sign "biological hazards."

↙ All work should be conducted in a BSC Class II, with HEPA filters.

- Special lab coats should be used in this laboratory. Ideally, these lab coats should have a different color than other lab coats used elsewhere.
- Instruments used in BSL-3 cannot be used elsewhere. You cannot share equipment with other laboratories, such as refrigerators, freezers, centrifuges, PCR, ELISA readers, etc.

BSL-4 laboratories: BSL-4 have the highest security levels. Only here are you allowed to work with microbes that may cause serious infections in people and can cause pandemics where little or no preventative measures are available nor a known cure. Microbes that belong to BSL-4 are all viruses such as the Ebola virus. Extensive specialized BSL-4 training is required before working in these laboratories and no one is permitted to work unless they have received hands-on training for BSL-4 work, with annual reviewal of laboratory safety plans and procedures. Advanced training in handling of biohazards is also required. In addition to the requirements listed for BSL-1, BSL-2, and BSL-3, following should be added:
- All work should be conducted in a BSC Class III, with HEPA filters.
- When working in a BSL-4 you need to wear full protective gear that covers everything, so no surface, skin nor hair is exposed.

Biological safety cabinets

In BSL-1, BSL-2, BSL-3, and BSL-4 laboratories, you should be able to find BSCs. These are cabinets designed to protect both the personnel and the environment from the microbes used inside these cabinets. These cabinets pull air in from below or from the air conditioning, filter the air into the hood, then they blow the air from the cabinets through HEPA filters out from the cabinet as sterile air, free of any microbes. The airflow inside the cabinet is designed in a manner so that the researcher never gets exposed to the air from inside the cabinet onto themselves. Therefore, there should be very little danger of getting infected while working. Specialized BSCs exist for each hazard level, such as BSC Class I, II, and III. Class II are available in five types A1, A2, B1, B2, and C1. The difference depends on the requirements and type of microbes being worked with, where they have different flow rates. Some have a negative air pressure to secure that there is no exposure to the air from the cabinet, Type B uses so-called single pass airflow, where the air does not mix nor recirculate. Type C has additional technical features. When working in a BSC the following procedures should be followed:

- ✓ Disinfect the working area before use. That way you prevent your samples from being contaminated.
- ✓ Place everything you need to use inside the BSC, before you start working, such as pipettes, pipette tips, flasks, etc.
- ✓ Allow the airflow to be turned on for at least 15 min before starting your work.
- ✓ Check the airflow.
- ✓ When starting your work, always go directly into the BSC. Avoid big hand motions, use only slow, intentional hand motions.
- ✓ Keep the samples inside the BSC.
- ✓ Arrange samples and other items inside the BSC so that nothing can hinder the airflow.
- ✓ Keep a trash bin inside the BSC and put all waste in it.
- ✓ When the work is over, disinfect everything before removing it from the BSC. You can disinfect the waste by spraying a disinfectant over it.
- ✓ Disinfect the whole surface area before you leave the BSC.
- ✓ Keep the hood turned on for 15 min after you finish your work inside it.

Vaccinations

Even though there is minimized risk of getting infected, if a vaccine exists for the disease you are working with, everyone who is working with that pathogen should get vaccinated. Working with biological human samples, they may contain viruses such as Hepatitis B, which is why it is recommended (or required) that hospital staff and other employees get immunized before starting their employment. Most institutions that work with microbes and human specimens recommend all students, researchers, and employees to be vaccinated against the following diseases: SARS-CoV2 (COVID-19), Hepatitis B, influenza, measles, mumps, rubella, and chicken pox.

Researchers working with soil samples, plants, and animals need to have an active tetanus vaccination (every 10 years).

Hepatitis C following needlestick injury[3]

After a 19-year-old nursing student was taking a blood sample from a 72-year-old female, she was recapping the needle, when she had a needlestick injury in her second finger of the right hand. The student was checked according to hospital protocol for hepatitis B and C antibodies and HIV antibodies immediately after the injury and after 3 and 6 months. 3 months after the injury she was hepatitis C positive! The student received appropriate treatment and after 36 months from the onset of the infection, she was declared cured.

Never be ashamed if you have a needlestick, and never try to hide it! All hospitals and laboratories should have protocols that should be followed in case of injuries like this one.

[3] This case was extracted from: Scaggiante, R.; Chemello, L.; Rinaldi, R.; Bartolucci, G.B.; Trevisan, A.; Acute hepatitis C virus infection in a nurse trainee following a needlestick injury. *World J. Gastroenterol.* 2013, *28* (4), 581–585.

CHAPTER 6

Radioactive materials

Introduction

Before working with radioactive compounds (Fig. 6.1), one must take required training or specialized courses in handling these compounds. Radioactive compounds emit radiation that is invisible. Some of these radiations are harmless, whereas other may cause serious illness such as cancer, damage to fetuses, or even death. Therefore, both laboratories where people are working with radioactive materials, and the radioactive materials themselves, are under strict monitoring by the authorities. Since it is not possible to see the danger and because many radioactive materials decay rapidly, one must know how to work with these compounds, approach them, and plan the work. In order to monitor radiation exposure, work with these materials often requires the people working in these laboratories, and even nearby laboratories, to wear personal radiation detectors. These may both be in order to monitor the *in situ* exposure and/or monitor the cumulative exposure over time.

Figure 6.1 The radioactivity hazard sign.

The laboratory

Designing a laboratory for work with radioactive materials depends on the type and the quantity of radiation to be allowed to be stored inside the

Handbook for Laboratory Safety
ISBN 978-0-323-99320-3
https://doi.org/10.1016/B978-0-323-99320-3.00014-8

room. Radioactive materials are categorized into four hazard classes, depending on how hazardous they are to human health. Likewise, laboratories are categorized into three hazard groups, called A, B, and C, where those working with low-risk radiation do that in laboratories class C. The standards for the laboratory design such as the benchtop, drains, sinks, fume hoods, etc., and the conditions and personal protective equipment depends on whether you are working in a class A, B, or C laboratory. Specialized personal protection that should be available in all such laboratories are lead apron or lead cloth of some type, depending on the type of nuclei, amount, and radiation type that will be used in this laboratory.

The classification of A, B, and C laboratories is based on the amount of activity of the overall quantity of radioactive materials to be stored there, as shown in Table 6.1. This includes storage of sealed and unsealed materials. Each laboratory must have a logbook, where all incoming materials are registered as well as their radioactivity at the time of their placement into the lab. After that it is important to be able to calculate, at any time, the radioactivity inside the room using the following equation:

$$A_{\text{Bq}} = A_0 e^{-k_d t}$$

where A_{Bq} is the radioactivity at time t, A_0 is the radioactivity when the material arrived at time zero, k_d is the decay rate, and t is the time since the material arrived. The k_d may be calculated using the following equation, where $t_{\frac{1}{2}}$ is the half-life:

$$k_d = \frac{\ln 2}{t_{\frac{1}{2}}}$$

The units used are called becquerel or Bq, where one becquerel is the unit for reciprocal second (1 Bq $= 1$ s^{-1}), defined as the activity of a

Table 6.1 Classification of laboratories based on the class of radionuclides to be allowed at any time.

Radionucleotide class	Type C (MBq)	Type B	Type A (GBq)
		Laboratory type	
Class 1	<0.5	0.5 MBq–0.5 GBq	>0.5
Class 2	<5	5 MBq–5 GBq	>5
Class 3	<50	50 MBq–50 GBq	>50
Class 4	<500	500 MBq–500 GBq	>500

quantity of radioactive material in which one nucleus decays per second. So $1\ \mu s^{-1}$ represents 10^6 disintegrations per second. In Table 6.1 the prefixes are MBq or megabecquerel (10^6 Bq) and GBq or gigabecquerel (10^9 Bq). 1 MBq is also called 1 rutherford. Many laboratories still use the older unit, curie (Ci), which is based on the activity of 1 gram of radium-226. Curie is defined as $3.7 \times 10^{10}\ s^{-1}$, or 37 GBq. So, 1 Ci is equal to 37 GBq, 1 mCi is equal to 37 MBq, 1MBq is equal to 0.027 mCi.

Example

One month ago, you received 25 MBq, ^{125}I labeled insulin. At the moment this is the only radioactive material in your laboratory. You are asked to calculate the amount of radioactivity remaining in the lab in Bq and Ci.

The half-life of ^{125}I is 59.5 days, so $k_d = 0.0116$ days^{-1}. The remaining radioactivity in the laboratory is

$$A_{Bq} = 25e^{-0.0116 \times 30} = 17.7 MBq$$

There are 17.7 MBq remaining, one month later, which is equivalent to 0.78 mCi.

All these laboratories require efficient air-conditioning and a fume cupboard or other handling box suitable for the work to be carried out, as well as a detector that can be used to scan your working surface at the end of the day to make sure that no radioactive drops or dust may have dropped on the table or any other surfaces. The following are the requirements for the three laboratory types, beginning with C:

Type C

A type C laboratory is intended for the handling of low activities and should be labeled as such. The design is similar to a normal chemistry laboratory. The surfaces, especially working benches, should be made of materials impermeable to moisture, resistant to ordinary chemicals, and easy-to-clean. Limit the number of equipment allowed in this laboratory and keep a specialized waste bin for radioactive waste. Based on the nature of the work, it is important that radioactive substances cannot escape into the air through the ventilation. The flow rate of air in the work benches should be at least 0.5 m/s when the height of the opening is 30 cm.

Type B

Type B laboratories are designed for handling radioactive substances. In addition to the requirements for Type C laboratories, these laboratories must fulfill the following: (a) a space for changing clothes and store protective clothing; (b) the floors must be easy to wash and must extend about 10 cm up the walls; (c) the working benches should be strong enough to hold e.g. lead shield or other radiation shields to prevent radiation; (d) the washbasins must be easy to operate, e.g., with arms or feet, without touching anything; (e) there must be negative air pressure in the room, so radioactive substances cannot escape the room.

Type A

Type A laboratories are intended for large-scale use of radioactive substances. These laboratories require strict safety license with a detailed description of each nucleotide, safety plans, personal protective measures, and clear work procedure.

Radiopharmacy

There is one more type that needs to be mentioned, which is the radiopharmacy laboratory, where pharmacists are preparing radioactive nucleotides to be administered to patients or to be used for diagnostic purposes. The manufacturing and storage of these radiopharmaceuticals should comply with Good Radiopharmaceutical Practice. Normally, these laboratories follow type C or type B laboratories.

All these laboratories must set up guidelines to be followed:
- Access to the laboratory should be limited to trained personnel or students.
- Every employee or student that intends to work with radioactive materials shall have the appropriate personal protective equipment. This should include a buttoned lab coat, gloves, and other protections as necessary, such as a lead apron. The lab coat or the apron should not be washed with other clothes in case it has been contaminated with radioactive materials. In such case it might need to be disposed of.
- Ideally, the lab coats used with radioactive material should be of a different color than other lab coats. That reduces the probability of them being mixed with other laundry.
- The lab coats should never be worn or taken to the cafeterias or other places where people eat or drink.

↳ Gloves should always be worn when working with radioactive material. Change your gloves every time you suspect that they might have been contaminated. If you have an open wound or a cut and need to work with a radioactive source, it is better to postpone the work until the wound/cut is healed.

↳ Adsorbing material should be placed on the benchtop before starting to work with radioactive material. If any of the material spills down, it gets adsorbed. If no cover is available, it can be prepared by using a paper towel. These adsorbing materials have a thin plastic film underneath, preventing droplets from going through and onto the surface of the table.

↳ The energy of the radiation, not the amount of material, is the primary factor for determining the thickness of the protective equipment. If you are working with strong radiation sources such as ^{32}P and ^{33}P (hazard class 3), the work area should be warded of with something like a plexiglass. About 1 cm thick plexiglass should provide adequate protection against a radiation of 5 MBq (0.13 mCi) ^{32}P or 50 MBq (1.3 mCi) ^{14}C or ^{35}S. However, similar protection is not necessary when working with tritium (^{3}H).

↳ When working with easily dispersed radioactive material, such as volatile compounds or radioactive gas, it should always be done in a fume hood.

↳ You should never use or store food, drinks, tobacco, wallets, bags, cosmetics, napkins, or food/drink containers, not even a coffee mug, in laboratories where radioactive materials are used.

↳ There is a distinction between working with *open* and *sealed* radioactive sources. A *sealed* radioactive source is when the material is in a tightly sealed closed container, preventing it from being dispersed into the environment. An *open* radioactive source is when the material is in other forms, such as powder or radioactive liquids used in laboratories.

Risk factor and laboratory type

It matters how work is carried out, based on the isotope you are working with. This has been described mathematically as r_{work} or the *work factor* and describes the risk (X) that the work may contain. Simple work with solutions, such as dilute solutions, is given the factor 0.1. General chemistry work has the factor 1, where complex chemistry work or work with dry material has the factor 10. Work with volatile dry compounds has the factor 100. Type C laboratories should not exceed risk factor $X \leq 1$, where Type

B laboratories are allowed to have risk factor of $X \leq 1000$, but there is no maximum for Type A laboratories.

The risk factor is calculated according to the following:

$$\sum_i \frac{A_i}{A_{\text{ref}}} r_{\text{work}} < X$$

where A_i is the amount of activity of the radioactive compounds you are working with, and A_{ref} is the maximum amount permitted in the type of laboratory you are working (see limits in Table 6.1).

Example

You are evaluating one of your laboratories where there is complex chemistry work with 1 MBq of ^{131}I (class 2) and 20 MBq of ^{3}H (class 4). Occasionally they are also working with 20 MBq of ^{14}C (class 3) in dilute solutions. The laboratory is classified as a Type C laboratory. Calculate the risk when all three radio-nucleotides are in use:

$$\sum_i = \left(\frac{1}{5} 10 + \frac{20}{50} 0, 1 + \frac{20}{500} 10 \right) = 2.44$$

The risk factor for this laboratory is 2.44, which exceeds what is allowed for Type C.

Bookkeeping and monitoring

In all laboratories where radioactive materials are used, there should be a detailed bookkeeping of all the radioactive compounds that are present in the laboratory at any given time. It is important to record the type of isotope, its activity, date, and the name of the person(s) responsible. Normally each compound is connected to the person who received the permission to work with that material. Whenever some of the radioactive compound is used, the amount should be recorded. That way, it should always be possible to see how much radioactive material and how much of each isotope is present in the laboratory at any given time.

To properly monitor and measure potential radioactive contamination in the work area, it is important to have an appropriate radiation meter in all laboratories where people work with radioactive material. At the end of each workday and if there is any suspicion that radioactive contamination has taken place, the area should be screened with an appropriate equipment, such as a Geiger counter. All areas where people work with radioactive compounds should be screened regularly for possible contamination, or at least every 3 months. If there is any radioactive contamination observed in the area, the source must be found, and the area cleaned. The protocols must then be updated accordingly to prevent any further future contamination.

Risk of exposure

All radioactive material should be stored in a way that prevents radiation hazards. It requires that the material(s) must be placed directly back into storage after use. Likewise, it is important to keep stock solutions or other solutions with radioactive compounds away from your working area or the benchtops. Put them back into the refrigerator, freezer, or behind a shield (e.g., plexiglass) as soon as it has been used. All solutions that include any radioactive material should be labeled with the isotope, activity, date, and the name of the person responsible for the work. The strength of radiation should always be as small as possible. In storage areas, the radiation strength should not surpass 7.5 μSv/h. However, if it is stored where other employees are working, it should be limited to a maximum of 2.5 μSv/h.

In most countries, there is a requirement that the radiation exposure of every employee, student, and other people working in proximity to radioactive substances be monitored and recorded. Employees and students that use radioactive materials in their work should keep the radiation exposure to less than 20 mSv/year, where the eye lens, skin (1 cm^2), and other body parts should not exceed a radiation exposure corresponding to 150, 500, and 500 mSv/year, respectively. A person at risk of being exposed to more than $\geq$6 mSv/year should carry a special radiation monitor whenever working with radioactive material or in an area where they might be exposed to radiation.

The unit Sievert (Sv) is used to measure the levels of ionizing radiation on the human body and the level of radiation that a person is exposed to. It can be the amount of radiation absorbed by a specific body part, or over the entire body. The dose is usually measured as mSv/year, but an older unit is called rem, where 1 Sv is equivalent to 100 rem (roentgen equivalent man). A normal person should generally not be exposed to more than 1 mSv/year. However, if you undergo a CT scan you are exposed to more radiation than 1 mSv, as well as when flying in airplanes, due to cosmic rays.

Cosmic rays are the largest source of radiation for most people on our planet. If your work includes working with ionizing radiation, you may be allowed to be exposed to 20 or 50 mSv/year, in Europe or the United States, respectively. Sievert is not used for high dosage of radiation. High doses can cause acute tissue damage or in more serious case radiation syndrome. To evaluate exposure to high dose radiation another value is used called *gray* (Gy), which is defined as the absorption of one joule of radiation energy per kilogram of matter. The toxicity limit is 4 Gy.

Sv is calculated from $w_R \times D$, where D is the dose in Gy, calculated as J/kg, and w_R is a dimensionless factor (quality factor). For x-rays, gamma, and beta radiation, the w_R is 1. For alpha radions the w_R is 20. This quality factor, $w_R = 1$, reflects full body exposure. However, if there is only an exposure to a part of the body, w_R will reflect that organ exposure such as thorax, thyroid, and urine bladder $= 0.05$; bone marrow, colon, liver, lungs, and stomach $= 0.12$; skin $= 0.01$.

Example

You are working with ^{99m}Tc and you want to know how much radiation you are exposed to, during your work, if you stand about 1.5 m away from the source, for about 2.5 hours a day.

To calculate the radiation exposure, you need to start by looking at several factors. The intensity (I) of radiation decreases as the distance (d) is increased according to the inverse square law where:

$$I_1(d_1)^2 = I_2(d_2)^2$$

In this case you are working with an amount of ^{99m}Tc that constitutes 370 MBq, so the daily dose (D) you are exposed to can be calculated using the following equation:

$$D = \frac{\Gamma A t}{d^2}$$

where Γ is the gamma constant, A is the source activity, and t is the time spent around the radioactive source each day. The specific Γ-constant for ^{99m}Tc is 16.60×10^{-3} µSv at 1 m per MBq per hour. Mathematically, we can also present this as as 16.60×10^{-3} µSv·m²·h⁻¹·MBq⁻¹ (Table 6.2). Adding this information into the equation, we get:

$$D/day = \frac{16.60 \times 10^{-3} \ \mu Sv \ m^2 h^{-1} MBq^{-1} \times 370 \ MBq \times 2.5 \ h}{(1.5 \ m)^2} / day = 6.8 \ \mu Sv/day$$

Example—cont'd

Table 6.2 Gamma (Γ) ray constants in $\mu Sv \cdot m^2 \cdot h^{-1} \cdot MBq^{-1}$ ($\times 10^{-3}$) for selected radionucleotides at 1 meter distance[1].

Nucleotide	$t_{\frac{1}{2}}$	Γ-constant
^{40}K	1.28×10^9 y	21.92
^{42}K	12.4 h	37.13
^{52}Fe	8.3 h	112.75
^{59}Fe	44.5 d	171.17
^{57}Co	270.9 d	0.382
^{60}Co	5.27 y	354.35
^{67}Ga	78.3 h	25.03
^{68}Ga	68.2 min	148.51
^{99}Mo	66.0 h	21.29
^{99m}Tc	6.02 h	16.60
^{111}In	2.83 d	55.42
^{113m}In	102 min	39.59
^{123}I	13.2 h	57.68
^{125}I	60.14 d	36.15
^{131}I	8.04 d	57.88

[1]Extracted from Pillay M. Dosimetric Aspects. In: *Textbook of Radiopharmacy, Theory and Practice* 3rd Ed. Editor, C. B. Sampson.; Gordon and Breach Science Publisher, 1999.

Assuming you work 250 days/year, the yearly dose would be 1.7 mSv/year, which is below the annual exposure limit of 20 mSv/year, and is therefore acceptable.

As described in the example above, you will be exposed to radiation when working with radionucleotides. To limit the exposure, different types of shields are available. However, the choice of shield should depend on the type of radiation the radionucleotides you are working with emit: (a) Alpha particles (He) require very little shielding, even your clothes or a sheet of paper may be enough to absorb alpha particles; (b) Depending on the strength of the beta particles, most of radionucleotides that emit beta particles require about 0.6 cm (1/4 inch) thick plastic; and (c) Gamma- and x-rays, however, require lead, concrete, or a lot of water. The thickness depends on the energy of the gamma rays (usually expressed in MeV, mega electron voltage).

When selecting a shield, one must look at its thickness and its density. The thicker it is, the more it absorbs and the higher density, the greater is the shielding. Attenuation coefficients (μ) are available for different shielding materials for many radionucleotides. The attenuation constants represent the quantity of radiation being absorbed. The shield effect may be expressed mathematically using the following formula:

$$A = A_0 e^{-\mu\chi}$$

where A_0 is the radiation intensity (in Bq) of the radionucleotide without shielding, A is the radiation intensity after shielding, μ is the attenuation coefficient (in cm^{-1}), and χ is the thickness of the shield (in cm).

Example

In the previous example, you were working with 370 MBq ^{99m}Tc. You decide to set a 2.5 mm thick lead shield between the radiation source and yourself. Calculate the radiation intensity after shielding. The attenuation coefficient for ^{99m}Tc is 23 cm^{-1} when you use lead (Pb).

$$A = (370\ MBq) \times e^{(-23 \times 0.25)} = 1.2\ MBq$$

By adding a 2.5 mm lead shield, you will decrease your exposure of gamma rays to almost 1 MBq.

Radioactive compounds

Radioactive materials are classified by whether they emit alpha-radiation (helium), beta-radiation (electrons), or gamma-radiation (electromagnetic radiation). These materials are also classified in four hazard categories depending on how hazardous they are toward people and the environment. Group 4 is the least dangerous one. In Table 6.3 there is a list of common isotopes found in these groups.

Since radioactive materials emit different radiations with different strengths, the storage requirements may differ significantly. The criteria is that the storage of radioactive materials should be in such a way that as small amount of radiation as possible is emitted from the material during storage.

Table 6.3 Classification of the most commonly used radionucleotides.

Radiotoxicity class	Isotopes				
Class 1	Pb–210	Po–210	Ra–226	Ra–228	Ac–227
	Th–228	Th–230	U–232	Pu–238	Pu–240
	Pu–242	Am–241	Cm–244	Cf–252	
Class 2	Co–60	Ge–68	Sr–90	Ru–106	Ag–110m
	I–124	I–125	I–131	Cs–134	Ce–144
	Sm–151	Eu–152	Bi–210		
Class 3	C–14	Na–22	Na–24	P–32	P–33
	S–35	Cl–36	K–43	Ca–45	Sc–46
	Mn–54	Fe–52	Fe–55	Fe–59	Co–57
	Co–58	Ni–63	Cu–67	Zn–62	Zn–65
	Ga–67	Ga–72	As–73	As–76	Se–75
	Br–82	Rb–84	Rb–86	Sr–82	Sr–85
	Sr–89	Y–88	Y–90	Zr–95	Nb–95
	Mo–99	Ru–103	Pd–103	In–111	Sn–113
	Sb–124	Sb–125	Te–132	I–123	I–132
	Ba–133	Ba–140	La–140	Ce–141	Pm–147
	Sm–153	Gd–153	Ho–166	Tm–170	Yb–169
	Ta–182	W–185	W–187	W–188	Re–186
	Os–191	Ir–192	Au–198	Hg–197	Hg–203
	Tl–204	Pu–237			
Class 4	H–3	C–11	F–18	Cr–51	Mn–56
	Cu–64	Ga–68	Tc–99m	In–113m	Dy–165
	Pt–193	Tl–201			

If they emit weak radiation, that can be done by storing the radioactive material in a refrigerator, a freezer, or behind plexiglass. When they emit strong radiation they should be stored behind lead. The most dangerous radioactive nuclei should be stored behind concrete blocks covered with lead.

CHAPTER 7

Waste management

Waste disposal rules and regulations

A lot of the material that is used in laboratories is harmful for the environment, humans, and animals. Therefore, all waste needs to be sorted and disposed of appropriately. As a lab worker, you are responsible for sorting and disposing of the waste you produce in your lab work. Waste should be managed in a way to minimize potential danger from it.

Different areas (e.g., countries and states) have different rules and regulations about waste sorting and processing, so it is important that you familiarize yourself with the rules for your area and follow them. Furthermore, many institutions have their own additional guidelines regarding waste sorting. If you are not informed of the guidelines of your own institution during your initial training, ask about them. If you are not sure how to dispose of certain waste, ask a coworker or your supervisor for advice.

Labeling of hazardous waste

Clear and informative labeling is the key when it comes to lab waste. It can vary from location to location how exactly we want to describe the contents in our waste containers, but in general, the more accurately we know its contents the better. Then we are better able to respond if an accident happens involving the waste. The waste disposal can also be more affordable, as there is not the same need to take special precautions in case some of the contents require special handling.

In some areas, the labeling only contains information about what waste category is, or should be, in that container, while at other places it is necessary to define the approximate amount of everything in the waste collection unit (see Fig. 7.1). Again, the rules regarding this can vary quite a bit between locations and institutions, so make sure you know what is expected of you at your workplace.

Waste containers

When waste is disposed of, it is important to choose the right containers and label them correctly (see resistance of materials to different chemicals,

Handbook for Laboratory Safety
ISBN 978-0-323-99320-3
https://doi.org/10.1016/B978-0-323-99320-3.00004-5

113

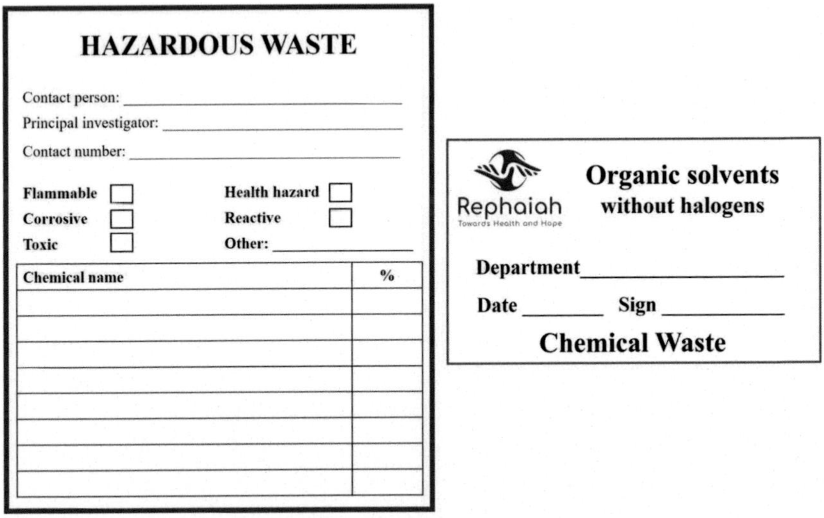

HAZARDOUS WASTE

Contact person: ____________________

Principal investigator: ____________________

Contact number: ____________________

Flammable	☐	Health hazard	☐
Corrosive	☐	Reactive	☐
Toxic	☐	Other: ________	

Chemical name	%

Rephaiah
Towards Health and Hope

Organic solvents without halogens

Department ____________________

Date ________ **Sign** ________

Chemical Waste

Figure 7.1 Examples of different waste labels.

in Table 4.6). If this is not done, the container can break with the resulting danger. Flasks that are used for hazardous waste should not be filled to more than 60%. If they are filled to a greater extent, a special safety cap should be used that can release potential pressure build-up in the flask. Hazardous chemicals should also never be stored where the sun can heat the content. That can lead to excessive pressure build-up that could lead to an explosion. Often waste might even have to be stored in a refrigerator or a freezer until it is completely disposed of. Do not use metal containers either since certain hazardous waste, such as acids, can eat through it. Waste containers, especially for liquid waste, should generally be stored in a secondary container (see Fig. 7.2). That way, if an accident happens, or the container is accidentally filled to overflowing, the waste is still contained in the secondary container, rather than becoming a larger chemical spill.

Exploding waste container

A few years ago, a hazardous waste flask exploded in one of the laboratories at the University of Iceland. The flask had been filled to the top and the cap had been screwed tightly on the flask. It was not kept in a secondary container.

During storage, an excess pressure was formed resulting in said explosion. The floor where the flask had been was damaged as a result as well. If the above-mentioned guidelines had been followed, and the flask had only been filled up to 60%, there had been less chance that this would have happened since the headspace could have better handled the storage pressure.

Figure 7.2 A waste flask stored in a secondary container. *(Credit: Benjamín Ragnar Sveinbjörnsson.)*

Broken glass should be placed in appropriate containers reserved for glass. Needles and other sharp waste should be put in appropriate sharp waste containers to minimize the chances of people stabbing themselves on it (see Fig. 7.3).

Waste from laboratories working with biohazards (microbes, blood and tissue samples, and animals) should be categorized by type before it is sent for incineration or landfill. Waste that includes radioactive compounds

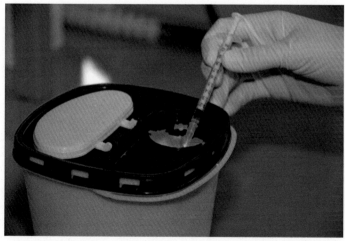

Figure 7.3 An example of a sharp waste container. *(Credit: Sveinbjörn Gizurarson).*

should be stored in a refrigerator or a freezer for at least 10 half-lives before it is disposed of (if possible). Special rules are also in place regarding radioactive waste that will be covered in more detail here below.

Sorting of hazardous chemicals

The chemicals we work with in lab are often toxic, health hazards, corrosive, flammable, carcinogenic, and generally hazardous in many ways. When it comes to their sorting, we want to minimize further hazards from these compounds and further exposure to them. Therefore, we need to consider their chemical behavior and reactivity so that their mixing does not result in a reaction or other interactions with other waste that might create additional hazards.

Section 13 of the SDS (see Table 4.1) contains information about disposal considerations, so make sure to read that section to be able to take appropriate precautions as necessary. When it comes to sorting of chemical waste, we generally start by separating it based on whether it is solid or liquid waste. Remember to label the containers clearly as to what type of waste they contain.

The chemical compatibilities that we need to consider tend to follow similar guidelines as chemical compatibility for storage, although we often have additional segregation when it comes to waste. Here are a few guidelines that may prove useful, based on the recommendations from the environmental protection agency in the United States:

- **Acids and bases** should not be mixed together in waste containers. They can react violently with each other and generate heat. In some cases, neutralization may be possible, allowing for easier disposal. This neutralization should be done carefully, as it in itself might be a violent reaction with heat and gases produced.
- **Alkali and alkaline metals, as well as aluminum, zinc, and other reactive metals/metal hydroxides** should not be mixed with any acids or bases in waste. The mixing of these can result in the generation of hydrogen gas, and can lead to fire and/or explosion.
- **Alcohols and water** should generally not be mixed with pure alkali or alkaline metals or metal hydrides. This could lead to violent reactions with explosive results. These should not be mixed with concentrated acids or bases either in the waste containers, as that can often generate a lot of heat. Chlorinated chemicals, especially chlorinating agents such as PCl_3 and $SOCl_2$, and other water-reactive waste should also

be segregated from alcohol/aqueous waste, as those can also lead to flammable or toxic gases, as well as fire, explosion, or heat generation.

↙ **Organic compounds and solvents** (e.g., aldehydes, halogenated hydrocarbons, nitrated hydrocarbons, and unsaturated hydrocarbons), especially reactive organic compounds, should be segregated from concentrated acids and bases, as well as from reactive metals, such as alkali and alkaline metals, aluminum, and zinc. Otherwise, it could lead to a violent reaction, fire, or explosions.

↙ **Cyanide and sulfide solutions** should not be mixed with acids, as their mixing could produce hydrogen cyanide or hydrogen sulfide gases, which are toxic.

↙ **Oxidizing agents** (e.g., chlorates, chlorites, hypochlorites, nitrates, perchlorates, permanganates, and peroxides) should be segregated from acids, both organic and mineral. These should also be disposed of separately from reactive metals, cyanide and sulfide solutions, and flammables in general. Oxidizers tend to increase the flammability of many of these compounds, and the mixing of these could lead to violent reactions, fire, or an explosion.

Accidental mixing of inorganic acids with organic solvents[1]

Nitric acid is a strong oxidizer that can react violently with many compounds, including organic compounds. The reaction time can also vary from minutes to days, which can create a false sense of security when no problems are observed upon mixing these together at first. Therefore, we need to learn from other's experiences in this regard.

In 2015, a student added some nitric acid wash into a glass waste bottle. The bottle already contained methanol and dimethylglyoxime. The bottle was then capped, and a reaction started taking place, leading to a pressure build up inside the flask. Eventually, when another student tried opening it a few days later, the bottle exploded, injuring three undergraduate students and a graduate teaching assistant at Texas Tech University.

Unfortunately, these types of accidents are too common, where waste containers explode or spray their contents due to incompatible chemicals having been mixed.

[1] This case was extracted from a *Lessons learned* document regarding the "Safe disposal of waste containing nitric acid" from Yale Environmental Health & Safety. Last accessed October 31st, 2021. https://ehs.yale.edu/sites/default/files/files/Lessons-Learned/nitric-acid-explosions.pdf.

In some places, halogenated waste is also segregated from nonhalogenated waste. The chemical waste is often burned eventually, and if the waste contains halogens, even if it is just a salt in a nonhalogenated solvent, then it requires extra treatment.

In general, make sure to read the SDS for disposal considerations, be mindful of the incompatibilities, and label the waste as accurately as you can, following the rules and regulations for your institute and your location. If the waste contains unknown chemicals, it becomes more expensive to have it disposed of, as it requires additional safety treatment for the disposal companies.

Disposal of biological samples and biohazards

Waste from laboratories where people work with biological hazards can be similarly diverse as waste from other laboratories. Biological samples, such as serum, saliva, biopsies, cell lines, and microbes, are all waste that can potentially be infectious and should be disposed of appropriately. All waste from laboratories where people work with biological samples, microbes, biohazards, and genetically mutated organisms should be sterilized using an autoclave, before disposed, securing no viable microbes or infectious agents in the samples. If there is no risk of infection from these samples, they can be disposed of as landfill or incinerated without special pretreatment.

Often, biological waste, such as blood samples, tissue samples, or biopsies are disposed in special bags (Fig. 7.4). In such cases these are bags, made of thick plastic material that are clearly labeled as biological waste. These are often stored in a freezer until they are taken, autoclaved, and disposed of via landfill or incineration. Waste that includes microbes or infected biopsies should always be disinfected and/or sterilized, e.g., by the means of an autoclave, before being disposed of.

Disposal of radioactive material

For radioactive waste there are international and domestic laws and regulations that apply. These apply to radioactive gases, solutions, solids, needles, glass, sharp objects, and biological samples. For radioactive waste, analogous *sorting* rules apply as for other waste. However, there should be a strict monitoring of the amount of radioactivity permissible to be disposed, on daily, weekly, or monthly basis.

Figure 7.4 Biological samples are placed in appropriate waste bag, before disposal. The Icelandic text on the bag translates to "Infectious material—to be incinerated." *(Credit: Sveinbjörn Gizurarson).*

Waste is not considered radioactive if the total activity of β- and γ-radiating nuclei is less than 10 kBq/kg and the total activity of α-radiating nuclei is less than 1 kBq/kg. A good frame of reference is also to not dispose of more than 10 kg at a time. Scintillation fluid is not considered radioactive if the solution contains less than 10 Bq/mL and has no α-emitting nuclei. Still, it is permissible to dispose of scintillations fluids that include the isotopes ^{3}H and ^{14}C, as long as their activity does not exceed 100 Bq/mL.

There can be differences in international, local (domestic), or institutional law and regulations regarding how much radioactive waste can be disposed of at a time, as well as how much cumulative radioactive waste is permitted to be disposed of per month or per year. Make sure to be informed about the regulations in your country as well as at your institution and follow them.

If the waste is no longer radioactive, remember to remove the label "radioactive material" that covered the container, before disposing of it. This also applies to waste that used to be radioactive, but has lost its radioactivity, or if the container is empty. Inaccurate labels found on waste, can cause confusion, fear, or discomfort among workers collecting the waste.

Many of the radioactive materials that are used in laboratories have a short lifetime. It is recommended that radioactive waste be stored in a refrigerator or a freezer until 10 half-lives have passed. During this time, the radioactivity is reduced to approximately 1/1,000 of what it was at the beginning. After 20 half-lives, the radioactivity is down to approximately 1/1,000,000. After that time, the material should have no radioactivity at all, and can be disposed of like any other waste. An example of that is ^{32}P, but it is recommended to store it for $14.4 \times 10 = 144$ days, in a lead container. After that time, it can be disposed of along with other chemical waste. This does not apply to isotopes with a very long half-life (e.g., years or decades).

Institutional waste disposal

Often the buildings have a dedicated location, where all the hazardous waste is collected in one place. This allows for easy collection from your institution to the appropriate disposal/incineration final destination. Some institutions take care of the final disposal/incineration of their institutional waste, while other institutions contract outside companies that offer hazardous waste disposal services. Under all circumstances, it is important that you know where (and how) you should leave your hazardous waste to be collected. Remember, when in doubt, consult your supervisor or a coworker.

CHAPTER 8

Chemical, biological, and radioactive spills

Introduction

Chemical spills happen, and it is probably the most common accident in laboratories. There is always a risk for a larger spill. Like for accidents, the first response is what matters. Small chemical spills can be cleaned up easily by employees and students, with little or no risk, but larger spills require training. For every spill, a risk assessment must be done before cleaning. Do you need personal protective equipment? People who will participate in the cleaning of large-scale spills should have documented training in cleaning spills and handling the waste.

After ensuring that no people are injured, one must start to minimize the further spread of the spill, followed by clean-up and waste disposal. For small spills, paper towels are the most important material to clean the area, as long as it will not react with the spilled material. Absorbents and/or neutralizing materials can also be used by sprinkling the materials over the spill before cleaning it up. There are different aspects we need to consider relating to spills. What is the nature of the compound? Is it liquid, aqueous, or an organic solvent, solid, semisolid, gas, vaporous, biological, or radioactive. Is it toxic, corrosive, contagious, or lethal? What happens if there are multiple compounds mixed in the same spill (Fig. 8.1)?

Companies, institutes, and universities should write protocols on how to handle different kinds of spills, depending on the size of the spill and location. Every employee must be trained to respond to spills. Laboratory managers should have hands-on training as well.

Handbook for Laboratory Safety
ISBN 978-0-323-99320-3
https://doi.org/10.1016/B978-0-323-99320-3.00005-7

Figure 8.1 When a fire in a laboratory had been extinguished, the firefighters entered the laboratory and found pools on the floor with unknown chemicals or a mixture of chemicals on the floor. *(Credit: The Greater Reykjavik Fire and Rescue Service.)*

Chemical containers[1]

Two different partially empty chemical containers were placed in the ordinary trash in two separate buildings. Then these containers were collected by the same trash vehicle and when these two containers came together it resulted in a fire. The fire caused the worker on the truck to be overcome with fumes, requiring emergency medical treatment. The nature of the fumes was unknown until later, when it was possible to retrieve the containers.

It is important to keep chemically contaminated containers, absorbents, paper towels, separate and do not mix them with other trash. Trash handlers may become very concerned when they see containers with chemical names or chemical hazard labels, such as a radioactive label.

[1] This case was extracted from: Furr, A.K. Emergencies (Chapter 2), in CRC Handbook of Laboratory Safety 5th Edition. CRC Press 2000.

First response to a chemical spill

Spills, small and large, can be dangerous. Strong acids and bases are corrosive and can cause serious chemical burns. Other chemicals, such as organic solvents, are easily absorbed through the skin, lungs, and other surfaces. Many compounds emit vapors that can be readily absorbed through the skin, lungs, eyes, and other surfaces. Therefore, you need to dress up. Cover your clothes/skin with appropriate chemically resistant and lightweight coveralls, wear goggles (not glasses), chemically resistant gloves, and if needed, respiratory protective gear. You must also cover your hair since you do not want the chemicals to be adsorbed to your hair.

Evacuation of the area is important to prevent others from being exposed to the materials. Minimize the distribution of the spill by the means of a chemical spill kit that contains necessary equipment to limit spreading. When the spill is not removed, it may continue to evaporate and get distributed into the air, which is why chemical spills must be cleaned up as soon as possible. In case you get part of the spill on yourself, make sure you flush with water for 15–30 minutes. Here below is the first response, when approaching a chemical spill and remember, never put your life in danger trying to play a hero.

There are a few things you need to consider first when approaching a spill. One of those is, what is the size of the spill? Your organization should have a spill kit, to handle small- and medium-sized spills. If the spill kit you have access to cannot handle the size, or if you require help, call for assistance. What about the compound itself? How toxic is the compound? What happens if you get it on your skin, inhale it, or it touches your lips and you ingest it? Do you have the necessary personal protective equipment to protect yourself from possible vapor, dust, airborne droplets, or other possible exposure to your skin, eyes, or the lungs. All these factors must be evaluated before approaching the spill.

Laboratory spill kits

Every laboratory should have a spill kit. They are available as a readymade package, or you can make your own. The content should contain the following:

Personal protective equipment

- Goggles (and face shield). Safety glasses do not give enough protection.
- Heavy neoprene gloves or other gloves depending on the type and nature of the chemicals in the laboratory.

↙ Disposable coveralls and/or corrosive apron that also covers the hair.
↙ Protective respiratory equipment.
↙ Chemically resistant boots (such as vinyl boots) or shoe covers.

Preventive measures

↙ Absorbent socks, to be placed around the edges of the spill, to prevent further spreading.
↙ Universal spill absorbent such as granulates, universal spill pillows, or absorbent pads, containing inert absorbents.
↙ Neutralizers may be important such as
  o Acid spill neutralizers: sodium bicarbonate, sodium carbonate, or calcium carbonate.
  o Base (alkali) neutralizer: sodium bisulfate.
  o Bromine neutralizer: sodium thiosulfate (5% solution).
↙ Mercury kit such as mercury decontaminating powder.

Tools

↙ Plastic dustpan and scoop.
↙ Disposable bags.

Efficiency of absorbing pillows[2]

In 1980, the Journal of Chemical Education published an article on spill control. Here they talked about preparedness and the importance that everyone working in a laboratory knows how to respond to spills. They showed that a bag or a pillow, containing 30 g of amorphous silicate, was able to absorb 250 mL liquid. Bags containing 120 and 480 g were able to absorb 1000 mL and 4 L, respectively.

They also tested different types of liquids and placed 30 g pillows on 250 mL water, nitric acid, and organic solvent. After 30 seconds, the remains of the spill were 3.2, 4.0, and 3.0 mL, respectively.

Absorbents should be available in every laboratory.

[2] This case was extracted from: Renfrem, M.M. Spill control in the laboratory. J. Chem. Ed. 1980, 57(5), A163—A166.

Response to chemical spill

↙ The role of the first responder is to try to minimize the distribution of the spill using an absorbent sock, and to minimize the damage.

✔ Notify the relevant people. Therefore, it is important that you ensure your own safety as well as warn others of the potential danger as needed.

✔ The first responder should stay on location and prevent further spread of the chemical spill, if possible. If you know what the chemical is, you should evaluate whether you can clean up the spill yourself or if you need help from an emergency response unit.

✔ The decision about whether to seek help from others is influenced by factors such as how well do you know the compound? How dangerous is it and how much spilled? If you need help, seek that immediately.

✔ If you do not know what chemical was spilled, assume the worst, and seek help. Do not start cleaning up a chemical if you do not know what it is. Close off the area to prevent further spread, call for help, and remain in the area.

Next steps

✔ If you know what the chemical is, you should clean it up right away using appropriate measures.

✔ Prepare by putting on the appropriate, disposable, personal protective equipment.

✔ Be careful not to pollute other areas. Take your protective clothes off, before leaving the room, so you do not contaminate door knobs, water faucets, towels, etc. If needed, use double gloves.

In case of a liquid spill

✔ If it is a small spill (not harmful), wipe up the spill using paper towels.

✔ When you clean up the spills, it is recommended that you start from the edges of the spill and work toward the center to minimize further contamination.

✔ If it is a hazardous compound or a large spill, a chemical spill kit should be used, with absorbents such as inert absorbent granulates, pillows or pads, before cleaning the area with paper towels. Depending on the absorbent material being used, it may be easiest to clean up the absorbents using a dustpan and a brush.

✔ Dispose of all contaminated paper towels, granules, pillows, or pads in disposable bags.

In case of a solid material spill

- Clean up the spill using a dustpan and a brush without distributing the solid material more than it has already been spilled.
- To remove fine solid particles, the area should be wet with an appropriate solvent and wiped up in a similar manner as described here above for liquid spills.

Final clean up

- Go through all items that may have been contaminated and make sure they are clean, functioning, and validated.
- Depending on the spill, some equipment and items may need to be discarded.
- Remove all your protective clothing. If your clothes have been contaminated, remove them immediately, wash your skin, and dress in clean clothes. The contaminated clothing may need to be disposed of.
- All contaminated items such as paper towels used for cleaning should be placed in a disposable plastic bag.

Final actions

- Write a report.
- Work with your local safety committee to go through the report, look at procedures and make amendments if needed.
- Evaluate what went wrong and how such accidents can be prevented from happening again in the future.

Radioactive material

Chemical spills, where radioactive materials are involved, need to be cleaned by documented trained personnel only.

First responder

- Same response as for other spills, except here it is important to make sure that there is an access to a Geiger counter or other measures to evaluate possible radioactive contamination on tables, floors, walls, equipment, etc.

If liquid radioactive spill

- Use same measures as described above.
- Once cleaning is ongoing, measure the radiation regularly. Also test the cleaning equipment such as the paper towels.

- Measure the radiation regularly during the cleaning until the radiation is the same as for the general background radiation.
- If general background radiation cannot be achieved, take a *cold material* (nonradioactive material of the same type as the spilled source), pour it on the spilled area, and mix well with the spilled area. Then continue to clean.
- During cleaning, make sure that the area is well labeled as contaminated area.
- In some instances, you may need to remove the surface that the spill took place on.

If solid radioactive material

- Use same measures as described above.
- For fine particles, the area should be wet with the appropriate solvent and wiped up in a similar manner as described here above for liquid spills. If the radioactivity cannot be removed, use *cold material* and continue cleaning.
- Check the area thoroughly using a Geiger counter or similar equipment, until the radioactivity is similar to the general background radiation.

Final cleaning

- Use similar measures as above.
- Measure the radioactivity in the disposable bag and calculate the radioactivity per kg waste. Using information about the isotope and the half-life, it may be best to store the trash until the radioactivity is the same as background radiation.
- When the radioactivity has reached background radiation, remove all radioactive labels from the trash bags. If workers that collect trash see packaging materials labeled radioactive, they may not want to remove such a bag. Make sure spills and trash are labeled correctly.

Final actions

- Use the same measures as above for chemical spills.

Employer's responsibility failure[3]

In 1984, a laboratory technician employed at a nuclear fuel production facility, complained to her employer that fellow workers were failing to clean up radioactive spills. She even drew attention to a spill by marking it with red tape. Several days later, when the area had not been decontaminated, she protested to the company.

Although the area was then cleaned, the company reassigned the laboratory technician to a new, temporary job and later laid her off.

Spills, whether chemical, radioactive, or biological in nature, are a health hazard and can even be life-threatening. So never leave a spill uncleaned. It must be cleaned immediately by yourself or another professional. Never leave the spill to the cleaning personnel, unless they have been trained to handle spills, with the training having been properly documented.

[3] This case was extracted from: Hukill, C. Whistleblower. Month. Labor. Rev. 1990, 113(10), 41–42.

Biological hazards

Biological spills may contain infectious materials, blood, tissue samples, or microbials. Therefore, it may be important to start by spraying a disinfectant over the spilled area. Then, let it stand for approximately 20 minutes. Be careful not to contaminate any doorknobs, faucets, towels, etc., when leaving the room. Always use double gloves in these types of circumstances. Then clean the area using a paper towel, starting from the edges in and toward the center as described here above.

Paper towels, and other items used for the cleaning, should be placed in an appropriate bag for proper disposal. All waste should be sterilized using an autoclave before being disposed. Finally, clean all surfaces again with soap and water or alcohol and wash yourselves thoroughly, especially the hands and the face.

Training

Everyone working in chemical or biological laboratories should be trained in cleaning up chemical or biological spills from different surfaces, such as wood, linoleum, and stainless steel. Whereafter, you take a swab test to measure how clean the surface became. This is especially important for those working with radioactive materials.

What type of table materials do you have in your laboratory, and do you have access to a spill kit?

CHAPTER 9

Accidents

Introduction

Small accidents and spills are the most common mishaps you will be exposed to in laboratories. Also, when you are working with others, accidents will happen around you. You may even discover holes on the back of your lab coat when you are washing it. This sometimes happens when many students are working together in a laboratory. In case of an accident your response must always be calm, measured, and immediate. Seconds can be crucial. If the accident involves chemicals or infectious material, your immediate response is vital. In this section, you will be guided through how your response should be, followed by quick guidelines focusing on different organs that may get exposed to chemicals.

A rule of thumb is that if you are the first respondent, start by protecting yourself. If you need to detoxify yourself, it is best to use the most effective detoxifying compound available, *water*. Use it right away and as much of it as possible, or until you are sure that you have removed the chemical from the exposed area. Always call for help and try to get the injured one to a doctor as soon as possible.

Here below are simplified guidelines on what to do if you are a witness to an accident:

First respondent

✔ Secure the situation to prevent further accidents.
✔ Provide first aid and call for help.
✔ You may need to call the emergency number. Describe your exact location and be specific when describing the accident. Ensure that someone will be waiting at the door to open up for the emergency medical technicians (EMTs).
✔ Take care of the injured one until help arrives.
✔ If the manager or the supervisor cannot go with the injured one to the hospital, be willing to go with the EMTs.

Handbook for Laboratory Safety
ISBN 978-0-323-99320-3
https://doi.org/10.1016/B978-0-323-99320-3.00010-0
129

✔ Ensure that you have sufficient contact information (name, address, phone number) of the injured one.
✔ If the accident involves chemicals or infectious materials, it is important that you can provide information about the chemical(s) or the infectious material, that the injured one was working with.

Next steps

✔ Do not move anything more than is necessary to secure the location until the situation has been explored sufficiently.
✔ Write down the names of any eyewitnesses to the accident.

Final actions

✔ All accidents should be notified to the local safety committee. They should then prepare an accident report and provide it to the appropriate authorities and human resources as appropriate. The safety committee should go through what happened, what went wrong, and what can be done to improve the safety.
✔ In case there are chemicals or infectious materials involved, you should be able to contact your local *poison control center* that provides information and counseling regarding harmful compounds and toxins. If in doubt, always contact your local emergency services.

Examples of accidents

Unfortunately, there are too many examples of accidents that could have been prevented if the safety aspects had been taken properly care of. Here below are a few examples:

✔ A postdoc was working with concentrated sulfuric acid. A bit of the acid spilled on their latex glove and burned right away through the glove. The individual received a second-degree burn. If he had been wearing nitrile gloves, this would not have happened.
✔ A British scientist was measuring the pH in a 4 L flask with hazardous waste when the flask dropped on the floor and crashed. Subsequently, the scientist slipped on the slippery floor and hit his head on the floor. He was all covered with the waste and received ugly burn wounds from the chemicals in addition to serious eye damage.
✔ A young, 22-year-old, employee of a lab in the United States received a hepatitis B infection. She had been working with biological samples, when she got an unknown droplet in her eye while transporting materials/samples. This droplet infected her eye and 4 weeks later she was diagnosed. She was not wearing safety glasses.

The eyes

In case of an eye injury, you must act quickly and be determined. Make sure that the injured person is guided to the eye wash station or to the sink (if there is no eye wash station). Rinse the eyes under a stream of water. It is best to use a *soft* flow, so that the water flow itself does not cause further damage. If you have access to eye wash bottles, use them until you reach the eye wash station.

Always ask if the person is wearing lenses. If so, remove the lenses and continue to rinse the eye with a lot of water.

When we get something into our eyes, it is normal to squint or close our eyes and keep them closed. Therefore, it is important that you assist the injured one, in keeping their eyes open by positioning yourself behind the person and *force* their injured eye(s) open, as shown in Fig. 9.1. This applies especially to cases where they feel a stinging or burning sensation in their eyes caused by the chemical. When the eyes have been opened and you start rinsing the eyes under the soft flow of water, ask the injured one to roll their eyes, and make sure to rinse the skin around the eyes.

When the eyes are at risk, always call for help and rinse the eyes until help arrives. Do not try to neutralize or detoxify the eye with an acid or base. It is important to rinse until a doctor or the EMT can take over. Continue, even if it takes up to 20 minutes before help arrives. If the injured one is wearing lenses, it is important to remove them immediately, because chemicals such as organic solvents can be trapped between the eye and the lenses, allowing the chemical to stay attached to the surface of the eye, causing further damage. We do not take any chances when the eyes are at risk. If you do not know how to remove the lenses, ask for help! Do not postpone the removal of the lenses. Only in situations where a professional can inform you that the compound is harmless to the eye, are you allowed to stop rinsing the eye.

Some laboratories may have eyedrops that can act as local anesthetics. If the manager is trained in using them, they may use it, because it will make the rinsing easier under the flow of water.

If an employee or a student gets a piece of broken glass into the eye, the injured person should be brought to a doctor. Trying to remove the glass piece yourself could cause serious eye damage so it is recommended to cover the eye(s) with a wet washcloth and bring the injured one straight to a medical professional.

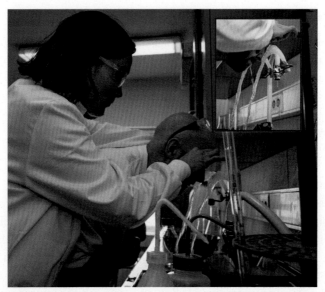

Figure 9.1 In case of an eye injury, make sure to bring the injured one to the eye washing station. Help them keep their eyes open so the waterflow can reach their eyes until help arrives. *(Credit: Birkir Eyþór Ásgeirsson.)*

After the injured person is under the care of medical professionals, clean up all the remaining pieces of glass from the floor and benchtops with a wet paper towel. Do not wait for someone else to do it! And be careful not to cut yourself on the glass shards yourself.

If you need to break some glass such as ampules, you should start by scratching the glass slightly (if it is not marked already) with a file, wrap a few paper towels around the area, and hold the thumb directly under the scratched area. Then you should break down and away from yourself.

If you need to remove a rubber hose that is tight on a piece of glass, you can sometimes start by trying to rotate it to get it looser. Be careful about using excessive force to remove the hose. It is often better to cut it off rather than try to force it off, since that has a greater risk of resulting in breaking of the glass.

Inhalation

Inhalation of hazardous compounds can be life-threatening. Therefore, it is important to react correctly and quickly. Call for help right away or have someone call the emergency services. Many gases are strong irritants such as

NO, NO_2, chlorine, and fumes from bromine, acids, bases, etc. Always make sure that the injured one is brought immediately to clean air, where they can breathe normally. Try to keep the injured one calm, but if they need help with the breathing, provide that. If you need to provide mouth-to-mouth breathing, take precautions, and consider possible contamination of yourself. If the person needs oxygen, provide that if available.

Never let the person lie down, not even in the locked position, but keep them in a sitting position. This is important because if inflammation occurs in the lungs such as after inhalation of corrosives, the inflammation will cause fluid to accumulate in the lungs. If the person is lying down, it will affect their breathing and may cause drowning, which is why it is important to stay in a sitting position. Even after arriving home, after being checked, inflammation is often delayed and may appear several hours later, when the person is gone to bed. Therefore, it is highly recommended that after an inhalation accident, the injured one should always have someone observing them for the next 24 hours.

Always consider your own safety and make sure that the air in the laboratory is safe. Never try to enter an area that may contain potential hazards. If there is a risk, make sure that only trained personnel in chemically resistant gear will go in first and ventilate the area. If your own safety becomes jeopardized, you may end up adding to the damage rather than helping the situation.

Since the harmful effects may not be apparent until many hours afterward, the injured one needs to get to a doctor as soon as possible, even if the incident did not seem to be serious. After the accident, the injured person should not ingest any alcohol or stimulants for 2–3 days after the accident.

If the person breathed cyanide (hydrogen cyanide, cyanogen or related compounds), an antidote should be administered as soon as possible. Let the injured one breath amyl nitrite[1] and bring them to fresh air and provide help with breathing if needed. Get them to a doctor as soon as possible.

Strong corrosive chemicals can cause life-threatening inflammations and fluid accumulation in the lungs, that can result in death. These symptoms may not appear until up to 24 hours after the accident. Various gases and dust can damage the lungs in a similar manner as acids and bases.

---

[1] Amyl nitrite should be accessible when working with cyanide. In case of an accident, break the amyl nitrite glass ampule and let the injured one breath it in. Use up to five glasses with a few minutes in between each one.

All chemicals that are irritants or corrosive can cause this type of inflammation or fluid accumulation in the lungs. When a person inhales hazardous gases, you may see strong reactions, such as coughing or difficulties in breathing, right away. After that, there can be a long period when the person feels alright. This can last from a few hours or up to 24 hours. But after that, fluid may start accumulating in the lungs, with the risk of drowning in their own fluids, internally.

Other extremely hazardous gases are carbon monoxide, hydrogen sulfide, and vapors of mercury.

Ingestion

Ingestion of hazardous compounds is not so common, but if someone ingests a hazardous compound, the mouth should be rinsed carefully, and possibly make sure that the injured one get a glass of water or milk to drink.

First response is often to get the person to vomit. However, that requires some guidelines. There are circumstances when it is desirable that the injured one vomits, when advised by a poison center, but this does not apply to all cases. It is, e.g., strictly forbidden to make a person vomit if they ingested a corrosive compound. Then the corrosive will burn the esophagus on its way down and will make that burning even worse on its way up again. Also never try to neutralize strong acids or bases inside the stomach. That will generate extreme heat, causing severe burns inside the person, or it will generate carbon dioxide (CO_2) causing even more damage than already has happened.

Also, do not make a person vomit if they have ingested organic solvents. When vomiting, there is a high risk of the solvent entering the lungs and causing a chemically induced pneumonia which could result in oxygen deficiency, inflammation, and death. It is always important to bring the patient to a doctor as soon as possible.

If the injured one starts vomiting by themselves, however, do not stop it. Always make sure you have access to the nearest poison center. Bring the injured one to a medical doctor as soon as possible. If the chemical is unknown, or there is a suspicion that the ingestion was part of a suicide attempt, it might be necessary to take a sample from the stomach for further analysis.

Skin contact

Always secure your own safety and use appropriate gloves. When a hazardous compound gets spilled on the skin, it should be immediately washed with a lot of water. If the spill covers a significant skin area, go directly under the emergency shower. Take off all contaminated clothing. Do not think about neutralizing the area, but call for help and have someone call the emergency number. Continue to wash until help arrives.

Only in cases when the compound is found to be harmless, is it acceptable to stop the washing. Lukewarm water is preferred, if possible. Never use hot water. Warm water can open up skin pores and increase absorption through the skin. Cold water, however, closes the pores which could result in the compounds getting trapped inside the pores.

Remove your clothes[2]

A truck was bringing chlorine to the chemical storage facilities in one of the swimming pools in eastern Iceland. Unfortunately, the truck driver poured acetic acid into the chlorine tank, resulting in the generation of chlorine gas. People working in the area realized immediately that something was wrong because they felt a burning sensation when breathing and they started coughing and some of them fell down. The area was evacuated immediately, and everyone was transported to the closest hospital.

Now, a few years later, one person is still severely affected by the chlorine gas. He was the only one who did not take a shower and change his clothing at the time of the accident. The chlorine gas got adsorbed to his clothes, resulting in continuous absorption of chlorine across the skin. If he had removed his clothes, he would have eliminated the chlorine source.

[2] This case was extracted from an Icelandic news story by Asmundsdottir S. Klórgasslys í sundlaug Eskifjarðar rakið til þess að ediksýru var fyrir mistök hellt í klórtank laugarinnar: Fór miklu betur en á horfðist (English translation: Chlorine gas accident in Eskifjord swimming pool traced to acetic acid being accidentally poured into the chlorine tank of the pool: Ended up better than it seemed). *Morgunbladid*, June 28th, 2006. Last accessed October 31st, 2021. https://www.mbl.is/frettir/innlent/2006/06/28/klorgasslys_i_sundlaug_eskifjardar_rakid_til_thess_/.

Acid burns tend to make ugly wounds and the skin starts peeling off. On the other hand, bases can enter deeper into the tissue underneath the skin and cause damage to the fatty layer. Therefore, it is important to rinse well, and for a long time. Do not try to use antidotes or detoxify the area unless there is an antidote designed for the compound you are working with. In case you are working with hydrofluoric acid (HF) or fluoride (F⁻) in some forms, you may need an antidote. Workplaces where you have these compounds should always have the antidotes readily available.[3]

If the etching is on the skin, close to joints, it should always be handled by a doctor, as the scarring on the skin might result in limitation of skin mobility in this area. In the case of acid/base spills, it is possible to check if the area has been rinsed sufficiently by using a litmus paper (pH paper) to check the pH of the area. Wait at least half a minute and measure the pH level. If it is <6 (for acids) or >8 (for bases), continue rinsing. Repeat the test after another minute.

Keep in mind that certain compounds, such as organic solvents and solutions containing sodium cyanide, can cross the skin without being noticed, so it is important to be vigilant and careful.

Burns and frostbites

Remember to protect yourself with appropriate gloves when working with extreme temperatures. If you get burned, the burned area should be cooled right away with a lot of running water. This can be done in a sink or under an emergency shower. Burned clothing should be removed as soon as possible. Do not hesitate to cut the clothing away from the area. Rinse with water until the sense of pain has faded out, even if it takes hours, or until a doctor takes over. Adjust the water to a comfortable temperature. Same response should be used for hot oil or fat.

In case of frostbites, use the same response but instead of cooling, use lukewarm water and do not stop until the pain is gone. Do not bind the area.

Needle punctures

In the event of a needle puncture, make sure to report it immediately. In all cases, wash the wound right away with soap and water and disinfect it with rubbing alcohol or 2% iodine solution. In some instances, it may be important to start an appropriate drug treatment right away, such as with

---

[3] If working with fluoride, use calcium gluconate gel or calcium gluconate solution. It is important to completely wash the area with such solution, using a wet towel, or a paper towel in a plastic bag. If the antidote is not available, use a lot of water to rinse.

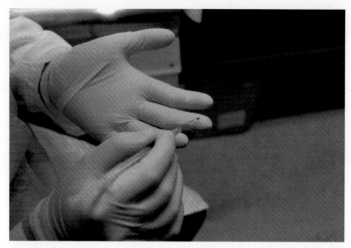

Figure 9.2 Needle punctures should be treated right away. *(Credit: Sveinbjörn Gizurarson.)*

antibiotics or antiviral drugs. In the event that you are working with blood or a biological sample, the sample together with the injured one should be taken to the nearest hospital, for human immunodeficiency virus and hepatitis C virus testing. This must be done immediately, in case there is a need to start necessary treatment *within 24 hours*. Needle punctures receive priority treatment, in health care (Fig. 9.2). If the needle contained organic compounds or solvent, you should also get the injured one to medical care as soon as possible. There are examples of even less than 100 µL of dichloromethane causing serious damage to a finger after a needle puncture. All accidents should be reported to the safety committee and the human resource department.

Emergency shower

The emergency shower should be used when the contaminated area is too large for a sink or if your clothes are contaminated. Also, if your clothes or your hair is on fire. Always provide help to the person involved in the accident and bring them under the emergency shower. Turn on the shower and remove all clothes that cover the contaminated or the burned area. In case of chemical spills, all clothes should be removed, to ensure that the compounds do not get trapped in the clothing. Trapped chemicals may otherwise continue to deliver chemicals across the skin.

Respect the injured one's personal privacy while helping them—this can be done by holding an extra lab coat up shielding the person if they had to strip down, until help arrives.

In case of fire, a fire blanket might be used as an alternative, at least if the emergency shower is far away. After the shower, help the injured one get clean dry clothing.

If someone turns on the emergency shower "for fun," that person should clean up after themselves. With that said, emergency showers should be tested regularly, but using the appropriate equipment.

First aid

It is recommended that everyone take a first aid course as well as a course in resuscitation, if possible. It is difficult to stand by, freeze, and not know what to do when a colleague is in danger. Studies have confirmed the importance of knowing the basics of how to provide first aid, while waiting for an ambulance or other specialized help.

CHAPTER 10

Fire safety and prevention

Introduction

Every employee and student must know the locations of the fire extinguishers in their building and the location of the emergency exits. It only takes seconds for an innocent spark to become an unmanageable fire. How you respond may distinguish if the fire is manageable or uncontrollable. There is a lot of fuel for a fire in all buildings, where papers and books are located, but then we have laboratories that may contain a large volume of chemicals, solvents, and even gases (Fig. 10.1).

Employees working in laboratory buildings must familiarize themselves with the fire prevention plans and the fire safety equipment provided in their workspace. Flammable liquids are one of the best fuels for fire. However, some chemicals may contain halogens or other compounds that can easily form life-threatening fumes and gases in the presence of heat and fire. In this section, we will discuss several types of fire extinguishers, response to fire, and evacuation plans.

Fire

Fire requires three things to start and live, oxygen, heat, and fuel. To extinguish a fire, we only need to remove or eliminate one of these three things. Fires are classified based on their fuel source. Similarly, the type of extinguisher is also based on the source of fuel feeding the fire. Here below is a description of the different fire classes, as well as the types of fire extinguishers.

Types of fire

Fires have been categorized into five to six classes depending on the source of fuel they are using and the hazards the fires can cause. There is a small variation in the classification, depending on geographical location. When

Handbook for Laboratory Safety
ISBN 978-0-323-99320-3
https://doi.org/10.1016/B978-0-323-99320-3.00012-4

Figure 10.1 Fire occurred in the chemistry laboratory at the University of Iceland on June 23rd, 1989. The figure shows the dense smoke that came forth when the fire fighters opened the window from the outside. Toxic fumes can easily develop under circumstances like this one, when different chemicals start reacting and burn together. *(Credit: The Greater Reykjavik Fire and Rescue Service).*

you look around at your work or inside your laboratory, evaluate what could be a fuel source, in case of a fire. Then, find out what type of fire extinguishers would be required to extinguish the fire. If you don't have a proper fire extinguisher at your work or in the neighboring rooms or the hallway, notify your safety officer, manager, or supervisor.

Class A (in Europe, United States and Australia) fire is when the fuel is paper, books, woods, clothes, and ordinary combustibles in general. This type of fire can be extinguished with water and film-forming foam fire extinguishers. Most office buildings have primarily these types of fires.

Class B (in Europe, United States and Australia) fire involves flammables and/or combustible liquids. The best fire extinguishers for this class of fire are carbon dioxide, dry chemical, and film-forming foam extinguishers. Film-forming foam extinguishers may only be used if applied correctly. Carbon dioxide extinguishers are often preferred since they do not leave as much of a mess to clean up as the others.

Class C (in Europe and Australia, but Class B in United States) fire is when flammable gases are the source. This type of fire is rarely manageable, but in case it is, dry chemical or carbon dioxide extinguishers are the option.

Class D (in Europe, United States and Australia) fire comes from combustible metals, such as magnesium, sodium, etc. *Never use water extinguishers or film-forming foam on this type of fire since that will likely result in an explosive reaction with the accompanying dangers.* Carbon dioxide extinguishers are not much good either. There are some class D—specific fire extinguishers that use class D powder. It may also be possible to use fine *sand* in case of metal fires. If you are working with combustible metals, make sure that you have sand or class D fire extinguishers close by, both where these chemicals are stored, and where you work with them. It is a good idea to take the appropriate precautionary measures in collaboration with the local fire department in your area as they need to know of this hazard in the event of an emergency.

Class E/K/C (Class E in Asia, Australia and previously in Europe, Class K in Europe and Class C in United States) are electrical fires. Electrical fires (up to 1000 watts) can be extinguished with dry chemicals or carbon dioxide *but not with water*. It is life-threatening to use a water extinguisher on electrical fires because of the conductivity of water. In buildings where there are sensitive, expensive electrical appliances and equipment, carbon dioxide is generally preferred over dry chemical since the powder is corrosive to electrical components and may damage the instrument.

Class F/K (Class F in Europe and Australia and Class K in United States) are fires where the fuel is cooking oil or fat. Film-forming foam extinguishers are the most suitable for this type of fire.

Peroxide forming chemicals

There are compounds known for developing explosive peroxides, even in diluted form. They are known to have caused serious explosions, and therefore it is not allowed to store them for longer than 3 months after breaking the seal (Table 10.1). When the seal has been broken, the inert gas, that is placed in the bottle, will be replaced by atmosphere, containing oxygen that can trigger the formation of peroxides.

Table 10.1 Compounds known for developing explosive peroxides.

Divinyl acetylene	Sodium amide
Isopropyl ether	Vinylidene chloride
Potassium (pure)	

The following compounds are able to develop peroxides, but only in their pure form (Table 10.2). Do not store them longer than 3 or 12 months after breaking the seal.

Table 10.2 Compounds able to develop peroxides when stored in their pure form. Here they are grouped based on their maximal recommended shelf-life.

3 months	12 months
Diethyl ether	Methylacetylene acetal
Tetrahydrofuran dioxane	Decahydronaphthalene glyme
Dicyclopentadiene dioxane	Tetrahydronaphthalene diglyme
Diacetylene tetrahydrofuran	Cyclohexane vinyl ether

Vinyl monomers generally develop harmless peroxides. These compounds can, however, degrade during storage, initiating a chain reaction forming combustible or explosive polymers (Table 10.3). Do not store these compounds for more than 12 months, unless they include sufficient concentration of the appropriate stabilizers/inhibitors.

Table 10.3 Compounds that may form combustible or explosive polymers during storage and should not be stored for more than 12 months.

Acrylic acid	Tetrafluoroethylene
Acrylonitrile	Vinyl acetate
Butadiene	Vinyl acetylene
Chloroprene	Vinyl chloride
Chlorotrifluoroethylene	Vinyl pyridine
Methyl methacrylate	Vinylidene chloride
Styrene	

Always make sure that you write on the package when the material was received, and when the seal is broken.

Compounds incompatible with water or air

Certain compounds are able to react with water or air, developing intense heat, fire, or even explosive fumes such as hydrogen. Examples of these are shown in Table 10.4.

Table 10.4 Compounds that are able to react violently with water or air.

Lithium (Li)	Sodium (Na)
Potassium (K)	Calcium (Ca)
Rubidium (Rb)	Cesium (Cs)

These compounds are dangerous not only in their pure form but also when they are in the form shown in Table 10.5.

Table 10.5 Compounds such as lithium, sodium, rubidium, potassium, calcium and cesium are dangerous in the following forms.

Hydride	Nitride
Sulfide	Carbide
Boride	Siliside
Telluride	Selenide
Arsenide	Phosphide

Nitrides and phosphides react with water and develop water reactive fumes, or even explosive hydrides. Some organic compounds like iso-cyanide, organometallics, acid chlorides, etc., react vigorously with water. Remember that acid anhydride, pure undiluted acids, and bases generate intense heat when diluted with water. They may even cause an explosion. If you are making an aqueous solution with these compounds, remember to measure the water first, then add the acid slowly to the water, and not the other way around.

In an event of a fire, do the following:

If you are the first person on the scene,

- ✓ Evaluate the situation.
- ✓ Call for help, push the fire alarm.
- ✓ Start evacuation.

Call the fire department

- ✓ When calling the fire department, give exact information about the location! At some institutions it is preferred to call the campus security/safety number and they contact the fire department immediately.
- ✓ Notify them about any injuries to people.
- ✓ Find information about the surrounding area, if needed, such as where chemicals are stored, and the estimated amount.
- ✓ Ensure that employees remove their cars from the area around the building, so there is free access for the firetrucks and ambulances.
- ✓ Ensure that someone can meet the firefighters upon arrival and provide them with any necessary information.

Too little information[1]

As previously mentioned, an accident took place at the University of California, Los Angeles, in 2008. A lab technician was working with *tert*-butyl lithium, when the material burst into flames and her clothes caught on fire.

When the firefighters heard that this was a chemical accident, they had to go to another location to pick up chemically resistant outfits. The time it took them to get the outfit and dress up was expensive. Unfortunately, the injured one died because of the accident.

There are several lessons to learn from this accident, one is the information provided to the fire department. The chemical was already gone, burst into the air, so even though it was a chemical accident, there was no need to spend precious time to get the outfit. If the information was presented correctly, they might have entered the laboratory sooner and saved the lab technician.

[1] Personal interview with the Safety Committee at UCLA.

Next steps

- ✔ Close all windows and doors, also in the neighboring buildings.
- ✔ If there is a lot of smoke or toxic fumes, make sure that you not only close the windows but also increase the heat in the rooms if you are in a neighboring room or building. That will expand the volume of the air, limiting smoke and fumes from entering.
- ✔ Make sure that all employees, students, and guests leave the area immediately and congregate at the designated fire assembly point.
- ✔ Continue to ensure that the ambulances, firetrucks, and the firefighters have clear access to the building (employees, students, and guests should move their cars away from the building to make it more accessible).
- ✔ If researchers or students in neighboring laboratories or buildings are working with flammable or toxic chemicals, stop the work (if possible).
- ✔ Do not try to be a *hero*! Do not risk your life. Only if the fire is manageable, should you start extinguishing the fire with the appropriate equipment.
- ✔ Notify your manager or supervisor about the fire.
- ✔ If there are rooms or areas that are locked, make sure that the keys are available, if needed.
- ✔ Have the blueprints ready, showing the location of fire extinguishers, storage of flammables, and other dangerous chemicals.
- ✔ Do not use elevators.

After the fire has been extinguished

⤙ Work with the safety committee at the company, institute, or the university. The committee should go over the sequence of events, look into what went wrong, what can be done to improve and prevent similar accidents in the future, and write up a report about what happened.

Preventative measures

Organic solvents, combustible metals, and other flammables or explosives are undoubtedly the most dangerous fuel in laboratories. The amount of these chemicals should be as little as possible in every laboratory. It is ideal to store liquids with flash points below 38°C in specialized refrigerators and to store at most 10 L of those solvents in laboratories that are about 30—50 m^2 (300—500 square feet). For solvents with flash points above 38°C, do not store more than about 20 L in a room of similar size as mentioned above.

Solvents, acids, bases, and oxidizing compounds should be stored in appropriate cabinets. Metals and other chemicals that can react with water and cause a potential explosion should only be stored in designated areas.

Fire extinguishers and fire alarms

There are several types of fire extinguishers and then there is a fire blanket. Working in a laboratory, you need to know the different types, when to use them, and which ones you might need and are closest by your work station. Learning to handle and use fire extinguishers and fire blankets should be obligatory for everyone, especially those working in laboratories. It makes us better prepared in the event of a fire and makes us likelier to remain calm and follow the proper procedure if we have tried it before. Employers should organize and offer training for everyone, regularly, e.g., in collaboration with the proper authorities.

Carbon dioxide (for fire class B, C, and electrical): Carbon dioxide extinguishers are common and one of the most important fire extinguishers for laboratories because it is mostly inert and does not leave the room in a mess. In laboratories where you have various solvents and flammable chemicals in particular, the carbon dioxide extinguisher should always be the first choice in the event of a fire in the lab, or where there are electrical appliances, equipment, and instruments. The carbon dioxide pushes the oxygen away, suffocating the fire. Additionally, it is so cold that it will cool everything down. Since carbon dioxide is a gas at room temperature, it does not leave any chemicals behind, which is especially good for instruments, as well as when cleaning the area after using it. However, carbon dioxide does not work against all fires and cannot extinguish embers.

Water hoses and water-based fire extinguishers (for fire class A): Water-based fire extinguishers are the most common as well as water hoses, especially in bigger buildings. The water hoses are often located close by the entrances, in hallways, and close by the emergency exits. Water extinguishers are not very common in laboratories since they are not feasible on Class B, C, or D fires. However, if there is a big fire that cannot be extinguished with carbon dioxide, large amounts of water should be used. Water has a two-fold effect on fire. It can cool it down, but during the cooling process it boils and forms a gas with 1700-fold increase in volume and that can push away the oxygen and suffocate the fire. Water-based fire extinguishers are especially common in office buildings.

Film-forming foam extinguisher, FFFE (for fire class A, B, and F/K): Film-forming foam extinguishers are powerful water-based extinguishers, where a surfactant is mixed in with the water. It is also called aqueous film-forming foam (AFFF). These surfactants enhance the extinguishing ability of the water greatly, especially for class A fires. FFFE can be found in many office buildings and laboratories, where there are no expensive and sensitive electrical instruments. The surfactants that are added to the water can damage electrical components exposed to these chemicals. This type of extinguisher makes a film between the fuel and the oxygen in the air, suffocating the fire. It also contains a large amount of water that can both cool down the fuel, and form a vapor that increases the volume of the water 1700-fold, pushing away the oxygen and suffocate the fire.

Film-Forming Fluoroprotein Foam (FFFP FOAM) extinguishers (for fire Class B) are alcohol-resistant extinguishers, effective on a wide range of organic solvents such as alcohols, where the film-forming foam seals the surface of the organic solvents, preventing oxygen from reaching the fuel.

Dry chemical fire extinguisher (for fire class A, B, and C): Dry chemical extinguishers can be found in many office buildings, where there are not a whole lot of electronics. These extinguishers are not suitable for laboratories since the powder is corrosive for electronic instruments and computers. When the extinguisher is used, it covers the fire with the chemical blocking the fuel from having access to oxygen, suffocating the fire.

Foam extinguisher (for fire class A, B, and F/K): Foam is an alternative to water and dry chemical extinguishers. In many countries this is one of the most common types used. It works well on combustible material as well as flammable oils and organic solvents. The foam forms a blanket on top of the surface of the fuel, removing access to oxygen. Foam is frequently used where flammable liquids are stored and used.

Wet chemical fire extinguisher (for fire class A and F/K): Wet chemical fire extinguishers contain potassium salts that are sprayed as a fine mist, that makes a soapy layer on the surface of the fuel. It is said to be the best fire extinguisher for fire involving fat or oil. However, it may develop toxic fumes, so the area must be highly ventilated. It is, therefore, mainly used where hot oil is used, such as in restaurants.

Fire blanket: You can extinguish all fire if you can reach it early enough! Fire blankets are, therefore, helpful if a fire comes up during your work. These extinguishers should not only be a part of every coffee break room but also be in laboratories. Every employer should provide training for employees and students in using fire blankets. Learning to approach a fire, overcoming the fear, and extinguishing a fire under the supervision of someone experienced is so important for future situations. Panic and wrong decisions are often seen from people who have never tried, but training will increase the chance of people remaining calm, thinking straight, and following guidelines (Fig. 10.2).

Blankets are good for small fires as they start. They are put on top of the fire to suffocate it by eliminating the fire's access to oxygen. When using fire blankets, it is good to hold the top corners with each hand and wrap it around your hands to protect them. The fire blanket should then be open in front of you, protecting your hands, face, and body. Holding the fire blanket like this, you should approach the fire and lay the blanket on top of it and let it lie there for a few moments until extinguished. Sometimes it can be good to lift up a corner and use a carbon dioxide extinguisher to help completely extinguish the fire.

Sand (for fire class D): Class D fires cannot be extinguished with regular fire extinguishers. Some metals can even stay on fire under water. If your laboratory contains a storage of combustible metals, it is required that you have fine grained sand or a Class D extinguisher nearby. Optimally, the sand should be placed above the metals, so in case it starts burning the sand will

Figure 10.2 Hands-on teaching on how to extinguish fire using a fire blanket. Here, everyone was obligated to learn to approach a fire and use a fire extinguisher. *(Credit: Sveinbjörn Gizurarson).*

automatically fall down and form a dense mass, like a cement, over and around the metals. In case there are also explosive chemicals stored in the same room, do not try to extinguish the fire, but evacuate the area right away.

Fire alarms: Most, if not all, buildings should have fire alarms distributed in such a way that it is never far to reach one. This is especially important at research institutes, companies, and universities. The locations of the fire alarms should be clear and well labeled. Often there is a small red flag above them, so their location is visible when looking up or down the hallway. In the event of a fire where the fire alarm does not go on, you should turn it on by breaking the glass and pushing the button.

Evacuation plans and exercises

Every company, institution, and university should have an evacuation plan and a fire assembly point, where everyone should gather in the event of an emergency. The purpose of evacuation plans and evacuation exercises is to prepare employees and students to respond correctly to emergencies such as fires. There should be a key person or persons in every building who have a special role in case of emergency, or as it is in some universities that the first on the scene becomes the key person, managing the evacuation and the next steps. These people must know what their role and responsibility is if the alarm system goes on and respond accordingly. Everyone else must follow their lead and commands. People in wheelchairs must go to the designated area in their building and wait for firefighters or others to carry them up or down to ground level, so they can meet the others at the fire assembly point.

When the alarm system in a building goes on, all fire doors should close automatically to delay and/or even prevent the further spread of fire, smoke, and toxic fumes in the building. Remember that using elevators when the fire alarm goes on is prohibited and everyone must know the fastest route to exit and go immediately to the fire assembly point.

Where is Dr. John?

The firefighters had dressed themselves in appropriate clothing so they could dive into a building filled with smoke. They were on their way to rescue Dr. John. At the same time, all the colleagues were assembled at the designated area in shock, because Dr. John was the only one missing from the building. They knew that he was inside when the fire alarm went on. The smoke divers were risking their lives in a building on fire to search for him, but he was nowhere to be found.

Later, when the fire was finally under control, Dr. John joined his group from another direction. When the alarm went on he decided to go home, since he lived close by, and have a cup of coffee, without notifying anyone. His colleagues had been in shock, because he was missing, and the firefighters had been risking their lives trying to find him inside an office building that was on fire.

The guidelines are for your safety and for everyone else's safety. Make sure you know the procedures, so we do not make people risk their lives if it is not necessary.

The role of the key person in each building should be the following (it can vary):

- Evacuate the building and in particular all laboratories.
- Direct people to the right escape route, in the event of an emergency.
- Check the control panel, where and why the alarm went on. It may be possible to locate the fire.
- Call for help, using the local emergency number.
- Provide information in the event of a false alarm or other error.
- Be able to provide first aid.
- Notify the safety committee and relevant authorities.
- Secure clear access for firetrucks, fire fighters, ambulances, and/or the police.
- Provide information to employees, students, and guests.
- Write up the event in the safety logbook.

Exercise:

Walk through your laboratory and take notes where the fire extinguishers are located, the emergency exits, and the fire assembly point. Check how long it will take you to get to the assembly point, from your workspace. Check the location of the smoke detectors, alarm, fire extinguishers, and the nearest water hose, and compare the fire extinguishers with the work carried out in your workspace. If you work with chemicals, do you have the right fire precautionary measures?

CHAPTER 11

Pregnancy and lactation

Introduction

Being pregnant does not necessitate that you need to stop all research, but it does require everything you do to be evaluated. That means risk assessment. Working in a laboratory, you must be aware that certain chemicals and materials may cause harm to the fetus or affect a lactating child.

If you are not familiar with the compounds you are working with, you must obtain information. Either through SDSs (see Chapter 4) or through databases. In case you cannot secure full protection for a woman who is pregnant or lactating, you should inform her to stop her lab work. Also, if you are working with new, unknown compounds or microbes, never take any risks. One should never take the risk that the compound or the microbe you are working with may affect the development or growth of your fetus or your baby! This also includes if someone else is working with compounds or microbes inside the lab. Often, we do not know all the chemicals in the lab that may result in fumes in the air, to be inhaled by those working in that environment. Risk assessment is therefore a very important part of the process that must be implemented when someone becomes pregnant. As soon as you become pregnant, inform your supervisor, so they can make the necessary precautions.

Fetal development and exposure

The first trimester (the first three months of pregnancy) is the most critical period for the fetus, because this is the time when organogenesis is ongoing. This is the time when cells representing each organ are formed, and if a substance affects this process, a malformation may occur. Compounds that may affect the translation of genes or alter patterns of gene expression, inhibit cell interactions, or block morphogenetic cell movements, may result in developmental abnormalities. This may happen at any time during the pregnancy.

Handbook for Laboratory Safety
ISBN 978-0-323-99320-3
https://doi.org/10.1016/B978-0-323-99320-3.00011-2

During the second and third trimesters, each organ is maturing, and their function and physiology becomes established. Affecting this period may affect the size and functionality of the organs. One can say that every second during fetal development has its purpose, we do not want to risk that something we are working with, or something that someone around us is working with, will affect fetal growth and maturation. The genes are continuously expressing building blocks resulting in interactions between cells that secure the functionality of each organ. This is like a clock that starts at a certain time, and when it is finished it shuts down again.

We may have access to information about certain compounds or microbes from animal studies, but the effect or lack of effects in animals does not necessarily translate to humans. It is, therefore, always important to carry out your own risk assessment. An important factor in this risk assessment is the timeline. Are we looking at one time exposure, or continuous exposure of small/large amounts over a long period of time? Is there an exposure to other compounds as well, that may facilitate the uptake of certain chemicals, or block the uptake? Some mixtures may have additive agonistic, synergistic, or antagonistic effects when people become exposed to them together, in a mixture. These considerations are important to include in the assessment.

A Swedish study on pregnant lab workers[1]

In 1984, a group of Swedish researchers analyzed the medical birth register of infants born in 1976 and identified those who were born to women coded as laboratory workers. They found that a higher than expected number of infants among pregnant lab workers either died neonatally and/or had congenital malformation. Since then, laboratory safety and the use of personal protective measures has improved significantly. Increased knowledge about occupational health and personal protective equipment will reduce the likelihood of becoming exposed to harmful compounds. Therefore, we must always show the utmost care when we allow a pregnant woman to work in a laboratory.

[1] This case was extracted from Ericson, A.; Källén, B.; Zetterström, R.; Erikson, M.; Westerholm, P. Delivery outcome of women working in laboratories during pregnancy. *Archives of Environmental Health* 1984, *39*(1), 5—10.

Exposure to chemicals and microbes

As a general guideline, pregnant women should never work with organic solvents, carcinogenic, teratogenic, highly toxic, or radioactive compounds, nor biological hazards. Certain procedures may be acceptable if they are carried out in highly ventilated cabinets or biological safety cabinets (BSCs), where there is no risk of exposure. The women must receive training, education, and guidelines, so they become familiar with all the procedures and become aware of all the risks involved (Fig. 11.1).

Supervisors are not allowed to take risks when a fetus is involved! Be aware that chemicals may be absorbed through the skin (even through clothing), through respiration, and mouth. Personal protection is therefore particularly important as well as the use of ventilated cabinets or BSCs.

Certain chemicals may cause birth defects, preterm death, preterm labor, genetic defects, cancer, toxic effects, or sterility. Even in seemingly insignificant amounts. Do not work in the laboratory when such compounds are used. Certain biohazards are toxic to the fetus and the pregnancy, such as *Toxoplasma gondii*; *Listeria monocytogenes*; cytomegalovirus; rubella, Zika, and many other microbes.

Radioactivity is also harmful for the pregnancy and the fetus. There are cases showing that if the mother is exposed to isotopes such as radioiodine, or even X-rays during pregnancy, it may have a serious outcome in their offspring(s).

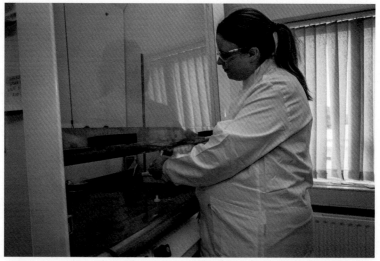

Figure 11.1 Working in a laboratory requires full personal protective measures and a workspace where you can eliminate the exposure to the chemicals you are working with. What PPE is missing in this picture? *(Credit: Birkir Eyþór Ásgeirsson.)*

Heavy lifting should not be a part of a pregnant woman's work description. Likewise, highly stressful work environment, such as high heat or cold, and even working under psychological stress, is not healthy for anyone, particularly pregnant and lactating women. If in doubt, talk with your supervisor or someone from your safety committee.

Soil contaminated with organic solvents[2]

In a study published in 2012, the prevalence of adverse birth outcomes was studied among mothers exposed to indoor air that was contaminated with soil vapor containing trichloroethylene and tetrachloroethylene. The study showed that pregnancies under these conditions were associated with higher incidence of cardiac defects. It also showed that trichloroethylene was associated with fetal growth restriction.

[2] The case was extracted from: Forand, S.P.; Lewis-Michl, E.L.; Gomez, M.I. Adverse birth outcomes and maternal exposure to trichloroethylene and tetrachloroethylene through soil vapor intrusion in New York State. *Environmental Health Perspectives* 2012, *120*(4), 616–621.

Exercise

Find an SDS for a compound or a microbe you are working with and look at the effects of this material on the pregnancy or the fetal development. Write down the risks, in case of exposure. Does your laboratory have ventilated cabinets or BSCs that will fully eliminate the risk of exposure?

CHAPTER 12

Occupational health and response plans

Introduction

Being an employer requires that you make sure that every employee arrives safely back home and that the work they do does not damage their health. That includes both physical and mental health. It requires you to make sure the work environment is safe, that everyone has access to the necessary personal protective equipment, and that the employees receive proper safety training for the instruments, chemicals, microbes, or environment they are supposed to work in. Such training must be documented. Universities must teach their students how to work inside a laboratory, what to look for, where the dangers are, etc., just like when you start driving. Afterward, the student should get a *driver's license* into the lab. We need to remember that although someone may have received basic safety training for lab work, giving them a license to work inside the lab, there are still compounds and environments that require extra knowledge or extra training before they should be allowed to work with those compounds or in those environments. Therefore, the employer's responsibility is to make sure that the employees have always received adequate training for their projects, so they can get safely back home after a good day at work.

Management and organizational structure

The managing director is ultimately responsible for all safety in the workplace. To make this work, they should set-up a well-structured central safety committee that works directly under them. Similarly, laboratory managers are responsible for making their laboratory a safe workplace. When you have a large workplace such as universities, there may need to be a network of safety committees that work in collaboration with the central safety committee (that operates directly under the managing director or

Error

 Handbook for Laboratory Safety
ISBN 978-0-323-99320-3
https://doi.org/10.1016/B978-0-323-99320-3.00001-X

president/rector of the university). Additionally, there may be a need to establish specialized safety committees that focus on different disciplines such as biosafety, chemical safety, radiation, fire precautions, human subjects, etc.

In universities, the professors, or researchers often do not want outsiders to intervene in their projects and refer to *academic freedom*. Not even faculty management! This would not be the case in institutions or companies. Sometimes, they do not want anyone to know what they are doing, especially if it relates to upcoming patent filings or new discoveries. This does not exclude them from fulfilling strict safety standards and securing that their students and personnel are fully knowledgeable about the risks in the work they do in the laboratory. In such cases, the local safety group should be aware of possible confidentiality issues and respect that, without lowering the requirements for high safety standards.

Every laboratory is unique, and the safety plan for each laboratory will, therefore, be different from all other plans. When the safety committee has designed a standard plan, each laboratory manager should be required to develop a written plan for their laboratory, based on the operations that occur in that laboratory. Unfortunately, many researchers find this work unnecessary and time consuming, because they want to focus on deadlines, goals, and are often under a lot of stress. These plans are, therefore, often neglected and missing in universities. When a researcher or a professor receives access to a laboratory facility, they need to understand that they will be responsible for the premises and should be required to write the necessary plans.

Are there any requirements for safety committee members? It is important that the members understand the work that will be conducted within different disciplines as well as the risks involved. The central safety committee will be involved in writing policies and general guidelines as well as providing the local safety committees with new regulations and information from relevant authorities. The central safety committee will also apply for necessary licenses and permissions on behalf of the workplace. It is also practical to involve someone who has human resource background in the safety committees, since the safety committees may need to implement new work procedures and guidelines in an environment where many researchers might just want to carry things out as usual.

I made a mistake! This is something many people find difficult to admit. Especially if a man made the mistake, and needs to admit it to a woman! In many cultures this would be painfully difficult. The safety committee, together with the management, must develop a psychologically safe environment, where admitting one's own mistakes is appreciated, so we can learn from them and make positive changes to the work environment.

The safety group(s) will need to oversee multiple things such as laboratory safety (different disciplines), waste disposal, spills, accidents, fire precautions, psychological safety, as well as building safety. In some work places the safety committee may need to be involved in medical programs, such as for field work.

Physical health

In 2014, there was an eruption in northern Iceland, where Holuhraun lava field is now. For geologists the nature is their laboratory, and this was a great opportunity to observe the eruption from a close distance. The problem, however, were all the gases that came up with the eruption.

After thorough consideration, the safety committee together with the head of the Faculty of Earth Sciences decided to require a health check on every employee and graduate student that would enter the volcanic area. Those having lung disease or limited mobility were not allowed to the area and everyone was required to carry an oxygen detector.

Reporting accidents

In order to learn properly from accidents at the workplace, it is important that employee's report all accidents. It is also good to have a clear pathway to report near-accidents, so that we can learn from close calls as well. For this to work effectively, a healthy safety culture needs to be present and constantly cultivated at the workplace, where people are willing and encouraged to talk openly about accidents without fear of repercussions because of it, even if it requires us to admit our own mistakes.

These accident reports should include a detailed description about what happened, including a description of the injuries that took place and how they came to be. Analysis should then take place to find the root of the problem, *why* the accident took place in the first place. There can be many factors at play here, e.g., faulty instruments, a moment's carelessness from the researcher, or unexpected reactions. List all relevant factors in the report. Then the key thing is to learn from the experience to minimize the chances of it happening again. If the accident happened during an experiment or a protocol that needs to be repeated, what could be done differently to reduce the chances of a similar accident repeating itself? This evaluation and the preceding root cause analysis is useful to do in collaboration with coworkers, where you can have a constructive discussion about how to enhance everybody's safety under analogous circumstances.

It is good if the workplace has a standard format for these accident reports that is easily accessible for all employees. It also needs to be clear where the accident reports should be submitted. Accident reports are of very limited use if no one ever reads them, so the relevant safety committees should review all of the accident reports regularly to be able to analyze the data. This applies especially to the situation where certain types of accidents might be unusually common. The appropriate steps should then be taken to reduce the odds of similar accidents repeating themselves even more. The safety committees are in a good position to distribute information about adjusted safety protocols when relevant, so this reviewal process is vital.

Healthy environment

Have you made a mistake? What happened when you told your supervisor, manager, or colleague about it? In a healthy workplace, we know that everyone makes mistakes now and then, and you are not afraid of admitting it because such information is appreciated and welcomed. We will also learn from it. Unfortunately, many are afraid! They believe that admitting mistakes makes you a lesser person than you are, they feel ashamed and may try to cover it up and hide it. Also, some managers reinforce such acts with their behavior, e.g., by becoming angry, freaking out, and/or scolding the person. In some cultures, it may even be difficult to admit a mistake to the opposite sex.

It is vital for a healthy company to make sure that the company is also psychologically healthy, meaning that everyone is respected and valued. It is important that every employee feel comfortable in sharing mistakes. We all know that they are sorry and will be more careful next time, but they also need to know that it does not affect how they are valued as employees.

A psychologically healthy environment ensures that you are valued and respected regardless of gender, race, sexual orientation, age, disability, health condition (e.g., pregnancy), ethnicity, color of skin, origin, religion, convictions, culture, or position. Discrimination or harassment, in any form, should never be allowed. There should be zero tolerance toward sexual harassment, inappropriate touching, and violence. Same with backbiting or slander. That will only poison the work environment and make some of the employees feel unwelcomed.

Risk management

Active risk assessment and risk management should take place at every workplace. This should be done both on macro scales, looking, e.g., at whole departments and the general type of work being done there, as well as on a micro scale for individual experiments (Fig. 12.1).

There are various risk assessment approaches in existence, and it might be useful for the safety committees to develop a standardized risk assessment form that is easy to fill in and not overly complicated. Too complicated forms run the risk of fewer using them properly, both from the perspective of filling them out and reading them. The PACE approach described in Chapter 1 is one approach that can be taken for risk assessment, especially on a micro scale.

In general, most risk management approaches, especially on a macro scale, follow the following four steps in one way or another:

(1) *Identify risks:* before we can manage the risks, we need to take time to identify them. Sometimes we can identify them because we know of the hazards associated with chemical classes being used or bacteria or

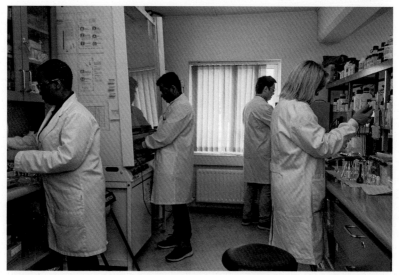

Figure 12.1 Work environment in a laboratory. *(Credit: Birkir Eyþór Ásgeirsson).*

viruses being studied, and sometimes we identify the risks from accidents that have happened. In any way, we need to start by identifying the risks.

(2) *Analyze/Assess risks:* the next step is analyzing or assessing the risk to see what can be done to minimize the danger from it. For chemicals, this often involves looking at information in the SDSs (Chapter 4) or reading about the hazards associated with the chemical hazard classes. For biological samples, this can involve reading up on what illnesses certain bacteria or viruses can cause, how serious they are, and whether cures exist.

(3) *Control risk/take action:* once we have assessed the risks, we need to take the appropriate steps to control it. This can include things like ensuring that the appropriate PPE and other safety equipment is accessible, providing additional training, and taking other preventative measures as necessary. It is also good to have a contingency plan ready for what to do in the event of an accident, and make sure that people are informed about how they should react.

(4) *Review and monitor:* this risk management is not really a four-step protocol that we can finish, but rather a continuous cycle (Fig. 12.2). We should continually review the risks, monitor them, and if our risk assessment changes, e.g., based on new information that we learn or after unexpected accidents that took place, then we need to readjust our controls.

Figure 12.2 A four-step risk management cycle. *(Credit: Benjamín Ragnar Sveinbjörnsson).*

Travel restrictions

Several years ago, a group of European students had planned to visit Thailand. The goal was to visit one of the universities as well as local biotech companies. One of the teachers was supposed to travel together with the group.

When planning the trip, there was a travel warning, published by the Ministry of Foreign Affairs (in Scandinavia, United Kingdom, and the United States), recommending people not to travel to Thailand that year, because of the situation being unstable at the time, with demonstrations and street violence.

When such a warning comes from the Ministry of Foreign Affairs, you should respect that. If you are responsible for the trip, you are responsible for making sure that everyone comes back safely. Make responsible decisions!

Emergency response plans

Most, if not all countries, have their own Department of Civil Protection and Emergency Management, that designs and publishes guidelines in case of natural disasters or catastrophes. Although all workplaces should have their own emergency plans, companies, institutes, and universities that have laboratories need to set up emergency response plans because of the nature of the operation within laboratories.

The remainder of this chapter will introduce ideas for appropriate measures in case of natural disasters or other hazardous situations that demand instant reaction, such as in the case of an explosion, earthquake, or a major chemical or biological accident with subsequent pollution.

Certain parts of these types of emergency response plans should be memorized. When danger arises, we tend to react illogically because of the fear that accompanies danger. By familiarizing yourself with the emergency response plan, you reduce the chances of illogical and wrong reactions.

We all have our "home building," that is the building where you have an office space, reading room and/or your research facility. So, familiarize yourself with your home building, and the buildings that you spend time in. Make sure you know where the emergency exits are and escape routes. Where is the assembly point in case of a fire or other emergency, and which doors close automatically if the fire alarm system goes on? How can I escape the building in the easiest and fastest manner? Do I need to use emergency steps or ladder? If you are an employee or have a managing role, you should also have access to or know the locations of the water intake and the fuse box.

In case of an emergency, your help may be needed to take care of people and bandage wounds. Having taken a first aid course, or other comparable courses, is therefore helpful. The more people who know the right reactions, the better chance we have of minimizing accidents from happening.

An explosion or a major chemical spill

An explosion sounds like something unrealistic or something that could never take place at our institute. But it is not so. Unfortunately, most universities that have a chemical laboratory have experienced an explosion in their laboratories. Thankfully, those examples are very few and often very small. However, these accidents and major chemical spills can happen without notice and take everyone by surprise. The best preventative measure to these accidents is to work in an organized manner and be well prepared every time you enter the laboratory.

When these types of accidents take place, the fire alarm should be immediately activated, the building evacuated, and everyone should gather at the fire assembly point. The only exception is in situations where the wind direction or other circumstances might jeopardize the safety where the fire assembly point is located, e.g., if toxic vapors may blow in that direction. Stay with your group until everyone has been accounted for or until the situation is cleared. Make sure people know that you are there, so no one initiates a search for you.

When the accident happens

- Keep calm.
- Call the fire department or the local emergency rescue services.
- Activate the fire alarm and evacuate the building. The fire alarm should close all fire-doors automatically and by doing that it limits possible pollution.
- Leave the laboratory right away, close the door, and get everyone to accompany you outside.
- Close as many doors as possible and windows, to limit pollution.
- After everyone has left the building, it should be closed.

After the accident

- The fire department has the necessary equipment to approach situations like these. They have trained individuals that have hazmat suits and are best trained to evaluate the situation on location. Never try to be a hero! No one should enter the building again, until the fire department has cleared the building for re-entry.

 ✔ If someone is injured, provide first aid and notify the emergency rescue team or the ambulance services. If you know of someone stuck inside the building, notify the fire fighters/emergency rescue team as well right away.

 ✔ The key person should be responsible for communicating with the fire-fighters, guiding them through the building (e.g., remotely through a headset), and provide information about where to turn off gas intakes, water, and electricity if needed.

Later, after the accident

 ✔ The safety committee needs to go over the situation, including the response plans and the reactions that were taken. Everyone injured should be followed up on. Buildings, instruments, and procedures should be reviewed, possibly together with the local authority.

Earthquakes

Earthquakes rarely give notice in advance. It is necessary to learn the right reaction and be able to execute them automatically. The best preventative measure to earthquake-related accidents is to make sure that cabinets and shelves are secured properly and fixed to the walls and/or the floors. Heavy items should also be secured to the walls or placed on the bottom of a cabinet or a shelf.

During an earthquake

 ✔ Keep calm.
 ✔ Evade furniture and other things that could start moving, like heavy items, books, and things that could fall out of shelves, ovens, broken glass, and possibly falling building parts.
 ✔ If you are in a laboratory, evade chemicals and lab equipment that could fall of the benchtop or out of shelves and break (Fig. 12.3). In case that happens, and many compounds get mixed, it is uncertain what could happen. Toxic compounds and vapors could develop. It is best to leave the laboratory right away and get everyone out with you.

Figure 12.3 On July 18th, 2007, the wall attachments in one of the laboratories of the University of Iceland gave in with the results that can be seen in the images. A PhD student who was working in the lab had just left when the accident took place. If he had been there a few seconds earlier, the cabinet would have fallen on him. *(Credit: Phatsawee Jansook).*

- If there are instruments turned on that can add to the potential danger of the situation, such as centrifuges and autoclaves, turn them off. Think first and foremost about your own safety and leave the laboratory as soon as possible, getting other people to leave the lab with you.
- Remember to DROP, COVER, and HOLD ON. You can do that in open doorways and hold on to the doorway itself but be careful of the door, it might slam. This can also be done in the corner of support walls or cover under a desk.

After the earthquake takes place

- If chemicals have fallen out of shelves in lab or if there is fire, make sure that no one is left behind in the laboratory.
- Afterward, close the facilities and call the fire department. They have the necessary equipment for situations like this. They also have trained individuals that have hazmat suits and are best trained to evaluate the

situation on location. No one should enter the building again until the fire department has cleared the building for re-entry.

- ✔ Check if anyone is injured. If there are injuries, provide first aid and call the emergency rescue team or the ambulance. If they cannot be reached by phone, identify the place of accident with a flag.
- ✔ Turn off the intake for gas and water and turn off the main electrical switch for the building if that is damaged.
- ✔ If you evaluate the situation as such that the structural integrity of the building has been jeopardized, or if you think the danger has not yet passed, the building should be evacuated.
- ✔ If there are chemicals that may have dropped from shelves, gas leak, etc., the building should be evacuated.
- ✔ You may need to turn on your national radio channels to follow possible guidance broadcasted by the National Department of Civil Protection and Emergency Management.

If the area needs to be evacuated because of an earthquake,

- ✔ Help all injured people to get out of the building.
- ✔ Notify everyone that the building must be evacuated.
- ✔ Leave the building through the nearest emergency exit.
- ✔ Gather at the fire assembly point and take down the names of everyone.

When the earthquake is over, and the situation is back to normal

- ✔ The safety committee needs to go over the situation, including the response plans and the decisions taken. Follow-up on everyone that was injured. Buildings, instruments, and response guidelines should be reviewed in collaboration with the local authorities.

Attacks and break-ins

If a situation arises where an individual poses a threat, is armed, intends to harm others or commit a break-in, it is important to keep calm and react correctly and with a level-head. Do not take risks with your own safety, the attacker could be armed. Contact police immediately and describe the situation. In an active shooter situation, the general guideline is "run, hide, fight." That is, if possible, run and escape the situation. If you cannot flee, hide, and turn off the sound of your phone. If hiding does not work, it may be necessary to fight the intruder as a last resort. Rarely, if ever, is it possible to talk reason into an active-shooter situation, so be careful about attempting that. Prioritize your safety! Also, if possible, warn others. Here below are a few things to keep in mind.

- ✔ Is the attacker on location? Is the attacker armed?
- ✔ Prioritize your own safety and the safety of others.
- ✔ Contact the police.
- ✔ When calling the police:
  - o Provide your name and where you are calling from.
  - o Describe the situation as accurately as you can.
  - o Describe the attacker as well as you can, appearance, such as clothing, identifying features, etc., their behavior: are they excited, calm, searching, how is their speech, are they throwing out threats, screaming, etc., and are they armed.
- ✔ Close the area from unnecessary traffic to prevent more people from being endangered and to prevent potential risky situations.
- ✔ Meet with the police, SWAT team, or other response teams if possible and provide them with answers.
- ✔ Think about crisis counseling and support for the victims and yourself.

Other dangers that can arise

Numerous unexpected situations can happen. Here below are a few situations that are good to be aware of:

Flooding and water leaks

- ✔ Turn off the water intake.
- ✔ Call the emergency response unit.
- ✔ Call the building supervisor or janitor if the incident takes place after hours or during the weekend.
- ✔ Minimize damage, by removing valuables from the area.
- ✔ Try to direct the stream of water to the nearest drains.
- ✔ Start clean-up.

Electrical outage

- ✔ Follow the emergency lighting to the nearest exit.
- ✔ Call the building supervisor or the janitor, if the incident takes place after hours or during the weekend.
- ✔ Do not try to fix the electrical board yourself.

⌐ Minimize damage as possible.
⌐ Keep in mind that there might be valuable samples in the refrigerators or the freezers. If the electricity is not back on within a certain time frame, it may be necessary to move the samples to other storage units. Storages storing unreplaceable or valuable samples should have a backup plan, regarding where samples can be relocated to.

Hurricane

⌐ Bind down lose items.
⌐ Make sure that all windows and doors are securely closed.
⌐ Cancel trips, lectures, meetings, and other commitments.
⌐ Do not jeopardize the safety of others.

Exercise:

Walk around in your laboratory and look for organic solvents, acids, bases, and other containers that may contain toxic or corrosive materials. Where are they stored? In case of an earthquake, could the bottles potentially fall on the floor? Suggest some preventative measures.

Index

Printed in the United States
by Baker & Taylor Publisher Services